FEMALE FIGHTERS

FEMALE FIGHTERS

WHY REBEL GROUPS RECRUIT WOMEN FOR WAR

Reed M. Wood

Columbia University Press
New York

Columbia University Press
Publishers Since 1893
New York Chichester, West Sussex
cup.columbia.edu

Library of Congress Cataloging-in-Publication Data

Names: Wood, Reed (Reed M.), author.
Title: Female fighters : why rebel groups recruit women for war / Reed M. Wood.
Description: New York : Columbia University Press, [2019] | Includes bibliographical references and index.
Identifiers: LCCN 2018060442 | ISBN 9780231192989 (cloth : alk. paper) | ISBN 9780231192996 (pbk. : alk. paper) | ISBN 9780231550093 (e-book)
Subjects: LCSH: Women and war. | Women revolutionaries. | Insurgency. | Women—Political activity.
Classification: LCC U21.75 .W67 2019 | DDC 355.2/3082—dc23
LC record available at https://lccn.loc.gov/2018060442

Columbia University Press books are printed on permanent and durable acid-free paper.
Printed in the United States of America

Cover design: Lisa Hamm
Cover image: © Newsha Tavakolian / Magnum Photos

Contents

Figures and Tables

Figures

Tables

Acknowledgments

As implied by the title, this book focuses on female fighters and the mobilization of women in war. The intention of this book is not to glorify female combatants or treat them as a novelty. Rather, it is intended to acknowledge the important contributions that many thousands of women have made to contemporary rebellions in every region of the globe, to better understand the conditions that facilitate or discourage women's participation in organized political violence, and to begin thinking about the possible ways in which the presence of female combatants might influence the groups and rebellions in which they participate.

The theoretical orientation of this book is toward the motivations and strategic calculi of rebel groups and their leaders with respect to the choice of incorporating women in the groups' combat forces or constraining and possibly eschewing their participation. Despite this focus on "elite" decision-making, I remain deeply interested in questions related to the motivations of individual women who elect to take up arms, how they navigate the social and institutional barriers that often impede or limit their participation in organized political violence, and their experiences within the (typically) overtly masculine organizations for whom they fight. Unfortunately, I am unable to address all of these questions in a single book; on the other hand, suspending consideration of the decisions made by the "rank and file" in order to more explicitly focus on group-level strategy formation allows me to develop a more focused, tractable set of arguments that help explain the variation in the prevalence of female combatants across armed groups and to begin exploring the strategic implications of their presence.

This book represents the culmination of a captivating intellectual journey. It reflects the intersection of numerous interests and fortuitous events. It is also the product of the invaluable input, insights, and collaboration of many different people. I officially began the project that ultimately led to the publication of this book in the summer of 2011, when I received as small seed grant to begin collecting data on female fighters. However, I trace the origins of the ideas and questions that formed the basis of this book to two unrelated trips to Central America years earlier. The first of these was a cultural education program in El Salvador I participated in during the summer of 2004. The organization that conducted the program was ideologically aligned with El Salvador's FMLN party, the party of the leftist rebels that had fought a twelve-year-long guerrilla war against the U.S.-backed government until reaching a negotiated settlement to the conflict in 1992. Among the most memorable aspects of this experience was the emphasis that the program placed on the roles Salvadoran women played in the rebellion and the number of female former FMLN guerrillas that worked with the program as guides, teachers, or volunteers. The second was a trip to El Salvador and Nicaragua to conduct predissertation fieldwork in the summer of 2008. While I was familiar with women's roles in the Salvadoran rebellion, prior to this trip, I was largely unaware of the substantial numbers of female fighters that had participated in both the Sandinista and Contra rebellions in neighboring Nicaragua. However, the handful of interviews I conducted with veterans groups in Managua, Esteli, and Somoto included a small number of former female guerrillas, and interviews with both men and women routinely referenced—at least in passing—women's roles in the armed struggles. Because the focus of my dissertation lay elsewhere, I treated these encounters simply as interesting anecdotes but largely irrelevant to the project at hand.

Only a few years later, while participating in a faculty working group at Arizona State University that focused on gender and human rights, did I reflect on these experiences and start to think seriously about patterns of women's participation in armed conflict. I am therefore grateful to the coordinators of this working group, Linell Cady, Carolyn Forbes, and Carolyn Warner, for inviting me to participate and for providing the small grant that officially launched the project that has evolved (over many years) into this book. I am also indebted to Patrick Kenney and the ASU Institute for Social Science Research for providing me additional funding that allowed me to begin the data collection for the project in earnest in 2012 as well as the School of Politics and Global Studies, which has provided additional resources since then. The Kroc Institute of International Peace Studies at the University of Notre Dame graciously provided me invaluable space, time, and resources to conduct research for this book during the fall of 2014.

Numerous graduate and undergraduate students have assisted in the data collection for this project over many years, and each deserves thanks. I would like to acknowledge the many hours of research conducted by ASU students Lindsey Allemang, Aaron Ardley, Carissa Cunningham, Jamie Dorer, Bryan Eddy, Cassandra Long, Rachel Olsen, Daniel Pout, Rahul Prasai, Justin Tran, Holly Williamson, and Oona Zachary. I also thank Michigan State University students Daniel Hansen and Zuhaib Mahmood. I am particularly indebted to Jacquelyn Schneider, who has contributed as much as anyone to collecting the information on women's participation in armed groups that served as the basis of the Women and Armed Rebellion Dataset (WARD), the dataset of female combatant participation used throughout this book.

This book is, at core, the product of many collaborative works, and two of my colleagues and co-authors deserve particular thanks and credit for their contributions. The first is Jakana Thomas, the co-creator of WARD. This dataset represents the intersection of independent data collection efforts that we had unknowingly undertaken simultaneously. The African rebellion cases included in WARD are largely based on previous data collection and coding efforts that she and Kanisha Bond undertook for their path-breaking study of women in armed groups, which was published in the *American Political Science Review*. Our previously published co-authored work on this topic was also highly influential in the development of this book. Particularly, chapter 1 builds on an earlier article we published in the *Journal of Peace Research* that investigated the role of group ideology in the recruitment of female combatants. Throughout this project, I have benefited tremendously from Jakana's thoughtful suggestions and critiques. I also owe a great deal of thanks to Devorah Manekin for her contribution to the experiment presented in chapter 5. The experiment discussed therein reflects a small part of a larger collaborative project on the topic of gender framing and audience attitudes toward foreign rebel groups. I thank her for allowing me to use a portion of that research in this book. Without either of their contributions this project would not have been possible.

Several additional people deserve thanks for providing me with information and access to the images of propaganda materials featured in chapter 2. I thank Lincoln Cushing of Docs Populi, Richard Knight of the African Activist Project at MSU, and Rick Sterling of the former Liberation Support Movement for their assistance. I also thank Meredith Lokken for sharing with me some of the images and sources she located in the process of her dissertation research.

A long list of other friends and colleagues have offered constructive critiques and suggestions that have improved this project over the years. Thanks to Marie Berry, Mia Bloom, Kanisha Bond, Govinda Clayton, Dara Cohen, Phoebe Donnelly, Richard Frank, Ben Goldsmith, Joshua Goldstein, Alexis Henshaw, Reyko Huang, Kelly Kadera, Sabrina Karim, Milli Lake, Cyanne Loyle, Zoe Marks, T. David Mason, Laura Sjoberg, and Meghan Stewart for comments on various portions and iterations of this project. I would like to thank the faculty and students of the Department of Peace and Conflict Research at Uppsala University, the School of Politics and International Relations at Australia National University, and the Department of Political Science at Rice University, and the Department of Political Science at Penn State University for providing me opportunities to present portions of this project and offering constructive feedback. The project has also benefited from the comments and intellectual discussions provided by the organizers and participants of multiple workshops on gender and conflict. These include "Measurement and Conceptualization of Female Combatants" at the School of International Service at American University, "Women in Violent Political Organizations" at the Radcliffe Institute for Advanced Studies at Harvard University, "How Gender Shapes Insurgency" at the School of Politics and Global Studies at ASU, and "Measuring Inclusion" at the Josef Korbel School of International Studies at the University of Denver.

The process of writing this book has been inextricably linked with my own personal journey. Throughout this process, I have benefited immeasurably from support, encouragement, and insights provided by Sarah Shair-Rosenfield. Lastly, Chiara, my own little rebel, has been an endless source of inspiration and motivation. I dedicate this book to them.

Abbreviations

ANC	African National Congress
ASG	Abu Sayyaf Group
BAAD	Big Allied and Dangerous Dataset
CFMAG	Committee for Freedom in Mozambique, Angola and Guine
CISPES	Committee in Solidarity with the People of El Salvador
CPN-M	Communist Party of Nepal-Maoist
DDR	disarmament, demobilization, and reintegration
ELF	Eritrean Liberation Front
EPLF	Eritrean People's Liberation Front
EPRLF	Eelam People's Revolutionary Liberation Front
ERP	Ejército Revolucionario del Pueblo (Revolutionary People's Army)
ETA	Euskadi Ta Askatasuna (Basque Homeland and Liberty)
FARC	Fuerzas Armadas Revolucionarias de Colombia (Revolutionary Armed Forces of Colombia)
FMLN	Frente Farabundo Martí para la Liberación Nacional (Farabundo Martí National Liberation Front)
FPL	Fuerzas Populares de Liberacíon Farabundo Martí (Popular Liberation Forces of Farabundo Martí)
FRELIMO	Frente de Libertação de Moçambique (Mozambique Liberation Front)

GAM	Gerakan Aceh Merdeka (Free Aceh Movement)
IO	intergovernmental organization
IPKF	Indian Peacekeeping Force
ISIS	Islamic State of Iraq and Syria
JSU	Juventudes Socialists Unificadas
KNU	Karen National Union
LNUP	Lahu National United Party
LSM	Liberation Support Movement
LTTE	Liberation Tigers of Tamil Eelam
MEK	Mujahadeen-e-Khalq (People's Mujahedin of Iran)
MFDC	Mouvement des Forces Démocratiques de Casamance (Movement of Democratic Forces of Casamance)
MILF	Moro Islamic Liberation Front
MK	uMkhonto we Sizwe (Spear of the Nation)
MODEL	Movement for Democracy in Liberia
MOSANAT	Mouvement du Salut National du Tchad (National Movement of Chad)
MPLA	Movimento Popular de Libertação de Angola (People's Movement for the Liberation of Angola)
NGO	nongovernmental organization
NLA	National Liberation Army (Yugoslavia)
NLF	National Liberation Front (Vietnam)
NAGS	Nonstate Armed Groups Dataset
NPA	New People's Army (Philippines)
NRA	National Resistance Army (Uganda)
NSA	nonstate actors
NSC	Namibia Support Committee
NUEW	National Union for Eritrean Women
OLS	ordinary least squares
OSPAAAL	Organization of Solidarity of the Peoples of Africa, Asia, and Latin America
PAIGC	Partido Africano da Independência da Guiné e Cabo Verde (African Party for the Independence of Guinea and Cape Verde)
PCS	Partido Communista Salvadoreño (Salvadoran Communist Party)
PFLP	Popular Front for the Liberation of Palestine
PIJ	Palestinian Islamic Jihad

PIRA	Provisional Irish Republican Army
PJAK	Partiya Jiyana Azad a Kurdistanê (Kurdistan Free Life Party)
PKK	Partiya Karkerên Kurdistanê (Kurdistan Workers' Party)
POLISARIO	The Frente Popular de Liberación de Saguía el Hamra y Río de Oro (Popular Front for the Liberation of Saguía el Hamra y Río de Oro)
PRIO	Peace Research Institute Oslo
PRTC	Partido Revolucionario de los Trabajadores Centroamericanos (Revolutionary Workers' Party of Central America)
RN	Resistencia Nacional (National Resistance)
RENAMO	Resistência Nacional Moçambicana (Mozambican National Resistance)
RPF	Rwandan Patriotic Front
RUF	Revolutionary United Front
START	National Consortium for the Study of Terrorism and Responses to Terrorism
SWAPO	South West African People's Liberation Organization
TELO	Tamil Eelam Liberation Organization
TOPs	Terrorist Organization Profiles
TPLF	Tigrayan People's Liberation Front
UCDP	Uppsala Conflict Data Program
UNITA	União Nacional para a Independência Total de Angola (National Union for the Total Independence of Angola)
URNG	Unidad Revolucionaria Nacional Guatemalteca (Guatemalan National Revolutionary Unity)
WARD	Women in Armed Rebellion Dataset
YJA-STAR	Yekîneyên Jinên Azad ên Star (Free Women Units) (PKK)
YPG	Yekîneyên Parastina Gel (People's Protection Units)
YPJ	Yekîneyên Parastina Jin (Kurdish Women's Protection Units)
ZANLA	Zimbabwe African National Liberation Army
ZANU	Zimbabwe African National Union
ZAPU	Zimbabwe African People's Union
ZIPRA	Zimbabwe People's Revolutionary Army

FEMALE FIGHTERS

Introduction

> To fight has always been the man's habit, not the woman's. Law and practice have developed that difference, whether innate or accidental. Scarcely a human being in the course of history has fallen to a woman's rifle.
>
> —VIRGINIA WOOLF, *THREE GUINEAS*

> I did not come to the front to die for the revolution with a dishcloth in my hand.
>
> —MANUELA, SPANISH *MILICIANA*

The image of a smiling young woman holding an assault rifle has become a common feature of media reports about the Partiya Karkerên Kurdistanê (PKK) and its allied Yekîneyên Parastina Gel (YPG) in Turkey and Syria. The extent of the female guerrillas' involvement in the conflict was dramatically illustrated during the Battle for Kobane in late 2014 and early 2015 as Kurdish militias—including thousands of female combatants—engaged in brutal fighting against the forces of the Islamic State of Iraq and Syria (ISIS) along the Turkish-Syrian border. The narrative of armed female warriors battling the forces of a radical Islamist army accused of mass rape, sexual slavery, and other atrocities against women (as well as men) made for evocative headlines, some of which boasted of the "badass" nature of these women and the "fear" they supposedly instilled in ISIS fighters (see Robson 2014; Salih 2014; Smith 2014). Yet, the sense of novelty inherent in most media reports on Kobane and other battles involving female Kurdish guerrillas belies the reality that women have often participated in organized political violence, including serving on the frontlines of many previous rebellions. While the women fighting on behalf of the various armed Kurdish factions in the region have perhaps received a disproportionate share of media attention in recent years, they are not unique.

While the direct participation of women in armed rebellion is uncommon in a comparative sense, numerous armed resistance movements fighting in different regions of the world have sent thousands of women into combat during contemporary civil wars, insurgencies, and terrorist campaigns. Across these groups, however, the scope and scale of women's participation varies substantially. Many estimates suggest that women made up at least a quarter of the combat forces of the Liberation Tigers of Tamil Eelam (LTTE), the Fuerzas Armadas Revolucionarias de Colombia (FARC), the Communist Party of Nepal-Maoist) (CPN-M), the Revolutionary United Front (RUF), the Eritrean People's Liberation Front (EPLF), and Frente Farabundo Martí para la Liberación Nacional (FMLN). Many other contemporary guerilla movements, including the Nicaraguan Contras, the Zimbabwe African National Liberation Army (ZANLA), the Provisional Irish Republican Army (PIRA), and the Popular Front for the Liberation of Palestine (PFLP), had smaller numbers of female fighters within their ranks. A larger number of terrorist and rebel groups have also sporadically employed female members as assassins, saboteurs, suicide bombers, or in other acts of clandestine violence.

Nor is women's participation in rebel groups (and the attention it generates in the media) strictly a phenomenon associated with contemporary armed rebellions. News reports from the time suggest that hundreds of women fought for Anarchist, Communist, and Republican forces during the Spanish Civil War (Lines 2012; Nash 1995) and female fighters were present in the Greek and Yugoslav Partisan forces during World War II and in the subsequent Greek Civil War (Batinic 2015; Poulos 2008). Thus, female combatants are neither a new phenomena nor one that is specific to any given culture, society, or world region.

Despite the participation of thousands of women in various armed resistance movements, rebellions, terrorist organizations, and guerrilla groups during the twentieth and twenty-first centuries, women represent only a small minority of the overall number of fighters involved in civil conflicts. Precise data are scant, but recent efforts to estimate women's participation in armed resistance movements suggest that fewer than a third deploy women in combat positions (Henshaw 2016a, 2017; Thomas and Bond 2015; Wood and Thomas 2017).[1] Consequently, while female fighters are not extraordinary per se, they nonetheless deviate from the historical norm. Most commonly, rebel combat forces—not to mention those of state militaries—comprise almost exclusively men. Women, to the extent that they are present in the organization at all, most commonly occupy positions in which they provide care, comfort, support, and encouragement to the male warriors: roles that generally mirror prevailing societal norms and cultural expectations.

The frequent absence of women from the frontlines of most rebellions is not surprising. The exclusion of women from organized political violence is a reflection of deeply embedded norms regarding men's and women's roles and duties in society. The battlefield is traditionally perceived as a "man's game," and military organizations exist as heavily gendered spaces that both encourage and reward masculine characteristics of aggression and dominance (Baaz and Stern 2009; Wood and Toppelberg 2017, 624). Women are therefore typically viewed as ill suited to the physical and psychological rigors of combat; moreover, their participation in war is often seen as inconsistent with—if not a direct challenge to—traditional military culture as well as prevailing social values. Indeed, as I explain in more detail below, deeply embedded gender norms and gender-based divisions of labor that ubiquitously link women to motherhood, innocence, and passivity and men to physical aggression and dominance appear to serve as the primary explanation for the limited participation of women in organized political violence (see J. Goldstein 2001).

While the biases of individual rebel leaders and prevailing societal norms often suppress the recruitment of female fighters, obedience to social norms and strict adherence to gender-based divisions of labor are often at odds with the realities of armed rebellion. Embracing these norms does not necessarily make for good guerrilla strategy. Given the asymmetric nature of most contemporary civil conflicts, rebels often find themselves poorly positioned to exclude willing recruits, regardless of their gender. Thus, incorporating women into the group's combat force arguably represents a rational response by rebel leaders to the strategic realities of armed rebellion against the (typically) militarily superior forces of the incumbent regime. The tension between embedded social norms/gendered beliefs and wartime resource pressures represents the dilemma inherent in rebels leaders' decision to recruit female fighters and deploy them in combat. Specifically, expanding the pool of potential recruits helps ameliorate rebel resource constrains. Yet, creating opportunities for women to directly participate in the production of violence often violates extant social norms and may represent a direct challenge to the social order—an outcome that can itself impose substantial costs on the movement.

I devote a significant portion of this book to understanding how rebel leaders navigate this decision and how they attempt to reconcile these competing forces. More broadly, I consider why some rebel movements ultimately incorporate women into their fighting forces while others persist in excluding them. In addition to investigating rebel leaders' motives for female recruitment, I also consider the strategic implications of this decision for the rebellion. Particularly, I investigate how the presence of female combatants can, under some circumstances, provide various benefits to the movements that elect to recruit them. These benefits include both the role that female combatants play in ameliorating resource demands, which I argue is a primary factor in their recruitment, but also a number of additional strategic benefits that armed resistance movements often appear to realize only once they have already begun to recruit female fighters into the their armed wing. These benefits largely center on the potential roles that female combatants (and their images) play in rebel efforts to secure international and domestic support for the group and its goals.

Summary of Main Arguments

The main arguments of the book are presented in two parts. I focus attention first on leaders' motives for female recruitment. In order to explain variations in the presence and prevalence of female combatants across rebel movements, I follow recent studies (e.g., Thomas and Bond 2015) that privilege "demand-side" explanations of women's participation in political violence. As such, the arguments developed in chapter 1 focus on understanding the conditions under which rebel leaders are willing to transgress or challenge existing gender norms by recruiting women into the armed wing of the organization and deploying them in combat roles. While acknowledging the importance of examining women's motives for joining an armed group, I assume that the leadership of the movement ultimately determines the recruitment strategy it adopts. In other words, I assume that rebel leaders act as gatekeepers in the recruitment process, and thus female fighters only exist when rebel leaders explicitly decide to train, arm, and deploy women in combat roles.

The overarching argument I make in the first chapter of the book is that rebel leaders are more likely to incorporate female fighters into the movement when they prioritize the potential strategic contributions of female fighters over the maintenance (or restoration) of traditional gender norms. I argue that the interaction of strategic and ideological factors shapes this decision. First, the severity of the resource demands faced by rebel leaders determines their "demand" for female fighters. As resource demands increase, such as during periods of intense fighting or following substantial troop losses, rebel groups become more likely to employ women in combat roles. Second, as previous studies demonstrate (e.g., Wood and Thomas 2017), the political ideology a group adopts strongly influences its baseline willingness to entertain the option of recruiting female combatants to meet resource demands. Leaders of rebel groups that seek to overturn existing social hierarchies and espouse gender equality (e.g., Marxist-oriented groups) are likely to be the least reluctant to employ female combatants and least sensitive to the potential costs of doing so, while those that seek to reinforce these norms (e.g., fundamentalist religious groups) are typically highly resistant to the notion of deploying women in combat because doing so is anathema to their beliefs and the values held by the constituents upon which the group relies for support.

While resource demands and ideology exert independent effects on the likelihood that rebel leadership opts to recruit female fighters, the latter exerts an important conditioning effect on the former. Specifically, while rebel groups espousing nonreligious ideologies become increasingly likely to recruit female combatants as conflict costs increase, fundamentalist and orthodox religious movements are comparatively less susceptible to these pressures. I argue that this reluctance results because the leaders of such movements perceive that any benefits resulting from the ability of female combatants to ameliorate resource constraints are offset by the severe costs associated with transgressing embedded social norms by deploying women in combat roles. Thus, even as resource demands mount, leaders of rebellions espousing fundamentalist religious ideologies generally refrain from recruiting female combatants. And even when they do recruit women for combat, they are likely to do so only in small numbers and to deploy them in only vary limited roles (e.g., as suicide bombers).

In the second chapter, I consider the strategic implications of rebel leaderships' decisions to recruit females. I begin by extending the strategic arguments outlined earlier in the book. Specifically, I argue that the decision to recruit female fighters results in a substantial expansion of the pool of potential fighters available to the movement. Consistent with the strategic logic that motivated the initial decision to recruit them, incorporating female combatants increases the number of troops available to the rebel group, thereby helping it to ameliorate its conflict-induced resource demands. I then consider additional strategic benefits potentially provided by the visible presence of women in the ranks. Drawing on previous research examining media coverage of female combatants and anecdotal evidence regarding the ways in which rebel groups utilize female combatants in their propaganda materials, I argue that rebel movements often highlight the presence of female fighters in their ranks as a strategy of domestic and international resource mobilization.

These benefits accrue through various mechanisms. First, the presence of female combatants spurs domestic recruitment by shaming reluctant young men into joining the rebellion and inducing sympathy among the local population. By a related mechanism, group efforts to highlight female combatants can induce resource mobilization among diaspora communities, who may feel guilty about "allowing" female combatants to fight a liberation war on their behalf. Second, because observers are more likely to associate men with aggression and violence and women with victimhood, peace, and compassion, the presence of female combatants is both novel and highly salient to domestic and international audiences. The novelty and salience of female fighters increases audience attention to the group and its goals. Additionally, these associations subtly but positively influence audience attitudes toward the movement. The presence of female fighters therefore represents one aspect of a group's efforts to construct a more positive narrative to counter government efforts to delegitimize it. To the extent these efforts are successful, rebel movements that employ female combatants and intentionally highlight their contributions to the cause may ultimately prove more effective at securing support from transnational advocacy networks and other external patrons. Success in mobilizing both domestic and international support is critical to the survival and success of a movement. Consequently, female fighters can play important roles in a group's ability to achieve its political and military objectives.

Definitions and Scope of the Arguments

The arguments I put forth here focus principally on the factors contributing to the prevalence of female fighters in civil conflicts and the impact of female combatants on the rebellions in which they fight. In order to develop a tractable theoretical argument relating to leaders' decisions to recruit female combatants I make several simplifying assumptions about "gender" and about the role of traditional gender norms in shaping actors' attitudes and behaviors. Before proceeding, it is therefore helpful to define several key terms used throughout the manuscript and define and discuss the scope (and limitations) of the primary arguments.

By *women* and *female* I mean individuals who are biologically female and possess female primary sexual characteristics. I therefore treat the biological sex of these individuals as a given trait that is observable to others within their society, most importantly by the leadership of the rebel organizations that might choose to recruit them. I acknowledge the important distinctions between sex and gender, the former being biologically determined and the latter being a social and cultural construction. Moreover, I recognize that gender is a complex, multidimensional concept and that there are inherent problems and limitations associated with assuming either sexual or gender binaries. A critical examination of the construction and meaning of gender within armed groups is a fascinating and important topic;[2] however, it falls outside of the scope of this project. Nonetheless, gender and gender norms play a critical role in shaping armed groups' and individuals' perspectives on the role of women in armed combat.

I use the terms *fighters* and *combatants* interchangeably. Herein, both refer to women to whom a rebel organization provides military training and weapons and whom they at least occasionally deploy in combat activities against enemy forces. This includes women employed in frontline combat, female suicide bombers, assassins, and female auxiliaries or members of civil defense forces who might be called upon to participate in combat when the situation demands. It excludes women who serve in other capacities such as fundraisers, recruiters, couriers, or informants but do not engage directly in organized violence on behalf of an organization. It also excludes women who spontaneously engage in acts of violence in support of the organization or its goals without the direct permission or authority of the group's leadership. I focus principally on combatants/fighters as opposed to women occupying support roles and other females associated with armed groups (camp attendants, "war wives," etc.)[3] for two primary reasons. First, the former category represents the most substantial and most observable deviation from the historical norm while the latter is much less conspicuous and more commonly encountered. Second, the motivations for rebels to deploy women in combat roles versus support roles are likely to substantially differ. Specifically, recruiting women to cook, care for wounded soldiers, or even carry weapons and war materials does not represent a fundamental challenge to traditional gender norms in the same way that training women to fight and sending them into battle with the intention of killing enemy troops generally does.

While I primarily seek to explain the prevalence of women in combat roles, it is important to note that women perform a number of important (often critical) duties within many rebel organizations. Individuals who serve as medical personnel, cooks, radio operators, truck drivers, weapons porters, fundraisers, recruiters, spies, couriers, safe house operators, and in other noncombat duties are just as important to the survival and success of an armed resistance movement as are the guerrilla fighters that directly engage in combat with enemy forces. Substantial numbers of women routinely serve in these integral noncombat support roles. Moreover, they are often considered to be formal members of the armed groups for which they fight, even if they never directly participate in the actual conduct of combat.[4] My intention is not to diminish the importance of these roles or to ignore the contributions made by the many women who serve in them. However, fully describing the various roles women play in armed groups and theorizing on the factors that influence variations in women's participation in these roles is beyond the scope of this book.

Finally, I define a *rebel movement* as an antigovernment opposition organization that uses armed force to attempt to achieve a stated set of political goals. These goals may include political reforms, the complete removal of the incumbent regime, greater autonomy for a geographic area or kinship group, complete independence from the central government, or other objectives.[5] In practical terms, the group must engage in armed confrontations with government forces that result in substantial numbers of fatalities.[6] This definition is broad and includes groups engaged in a variety of types of civil conflicts, including full-scale civil wars, rebellions, guerrilla wars, insurgencies, and terrorist campaigns. However, it excludes organizations engaged in strictly criminal activities that have no obvious political goals (e.g., criminal syndicates), violent groups that arise spontaneously and have no clear organizational structure (e.g., riots), groups engaged exclusively in violence against other nonstate groups (e.g., intracommunal violence and conflicts between rival militias), and organizations that do not engage in armed violence (e.g., nonviolent protest movements).

While this definition excludes many types of violent actors, these exclusions are necessary in order to develop a tractable and parsimonious theory of political violence. The different types of groups listed above have vastly different goals, interests, and organizational structures that complicate the development of a more general theory of women's participation in social violence. The focus on rebel movements facilitates theory construction and analysis; however, it should be seen as only a starting point for understanding the conditions that lead women to join violent movements.

Women's Participation in Armed Conflicts

Despite the comparatively limited attention the topic has received from scholars of political violence, space does not permit a thorough overview of the field's knowledge regarding women's participation in organized political violence.[7] However, several important findings and observations that inform the arguments presented in the subsequent chapters are worth highlighting in this introductory chapter. Additionally, questions regarding women's agency and their frequent exposure to bias and abuse in conflict environments are both normatively and theoretically important to any study of women's participation in armed groups. I therefore devote some attention to discussing these issues in this section.

Women and Political Violence

Women are routinely stereotyped as less violent than men and, as a result, frequently assumed to be the victims of violence rather than its perpetrators. Yet, there is ample evidence that women across a range of social and cultural contexts directly engage in wartime violence. Recent studies demonstrate that rather than simply representing victims of sexual violence during war, women are sometimes perpetrators of wartime rape (Cohen 2013b, 2016; Sjoberg 2016). Women have also directly participated in other atrocities, including genocide, terrorism, and other forms of intentional civilian victimization (Bloom 2011; Sharlach 1999; Sjoberg and Gentry 2007). Previous studies likewise illustrate women's willingness to participate in organized rebellion, occasionally in large numbers, when they are provided an opportunity to do so (Alison 2009; Coulter, Persson, and Utas 2008; Gonzalez-Perez 2008; Kampwirth 2002; Mason 1992). Finally, available evidence suggests that female combatants are a global phenomenon and sometimes emerge even within societies characterized by a high degree of patriarchy and the presence of rigid social norms (see Henshaw 2017; Thomas and Bond 2015; Wood and Thomas 2017).

Few previous studies have attempted to identify a general set of conditions conducive to the presence or absence of female fighters in rebel movements or to explain the remarkable variation in the prevalence of female combatants across armed groups. With few exceptions, existing research on this topic has focused on describing and explaining women's participation in specific rebel movements rather than attempting to understand variations in women's roles across groups. These studies provide important insights into women's participation in a small number of armed movements, but they do little to explain variations in the presence and prevalence of female combatants across groups.[8] More recently, a handful of cross-national analyses have begun to systematically investigate the factors that predict female participation in armed groups (e.g., Henshaw 2017; Thomas and Bond 2015; Wood and Thomas 2017). Taken together, the results of case analyses and recent cross-national research highlight a number of factors that help explain women's participation in rebel groups.

First, previous studies have highlighted the central role that prevailing gender-based divisions of labor and the depth of women's integration into social, economic, and political processes play in determining the likelihood of women's participation in armed resistance movements. Based on his extensive survey of literature, Joshua Goldstein (2001) concludes that deeply embedded gender norms offer perhaps the most powerful explanation of women's general exclusion from armed combat. Linda Reif (1986, 150) similarly asserts that the patriarchal attitudes used to legitimate women's relegation to domestic roles during peacetime also inhibit women's participation in guerrilla movements. More recent statistically analyses of women's participation in rebel movements also find that female fighters are more common among rebel movements in conflicts occurring in states in which women enjoyed greater prewar economic opportunities and social status (Thomas and Wood 2018). Importantly, there is also evidence that factors that weaken traditional gender norms—including those that are endogenous to the conflict—can create opportunities for women to participate in armed rebellions. David Mason (1992), for instance, argues that the combination of increasing economic vulnerability, rural-urban migration, and opportunities to engage with grassroots political movements facilitated women's participation in Central American revolutions. Consequently, while prewar gender norms appear to influence the odds of women's participation in armed rebellions, they are not immutable. Nor do they necessarily prevent women from joining rebel movements.

Previous studies have also linked women's participation in armed groups to the ideologies and objectives of the armed movements for which they fight. Like men, women appear to mobilize in support of a range of ideologies. However, existing studies find that groups advocating for economic redistribution (Henshaw 2016b, 2017), those that espouse gender equality (Thomas and Bond 2015), and rebel movements that embrace Marxist-oriented ideologies (Wood and Thomas 2017) are comparatively more likely to include female combatants. Case studies of the FMLN and Sandinista rebellions also suggest that the groups' Marxist agendas appealed to many women and encouraged their recruitment (Kampwirth 2002; Reif 1986; Viterna 2013). This is perhaps not surprising since challenging existing social hierarchies and advancing women's rights is often an important aspect of Marxist movement goals. Yet, support for fundamentalist Islamist ideology, which typically seeks to impose or reaffirm traditional gender-based divisions of labor, appears central to women's participation in movements such as Hamas, Al-Qaeda, and the Caucus Emirate (Chechen rebellion), including their decisions to become suicide bombers (Bloom 2011; Cragin and Daly 2009; Von Knop 2007).[9]

Similarly, despite the frequent association between nationalist ideology and patriarchal attitudes, armed groups espousing nationalist ideologies have successfully motivated many women to join armed their movement and to engage in violence on their behalf (Alison 2009; Hamilton 2007; Parashar 2011; Speckhard 2008). Thus, group advocacy for the advancement of women's rights and gender equality do not represent necessary conditions for female recruitment; nor is their absence an inherent deterrent to women's support for an armed movement. As I discuss more thoroughly in chapter 1, ideology is indeed central to a group's decision to utilize female combatants because it shapes both the leaders' individual perspectives on the appropriate wartime roles of women as well as their willingness to transgress local gender norms by recruiting women. Particularly, it shapes rebel leaders' perceptions of the viability of recruiting women as a way to manage acute human resource constraints.

An often overlooked but arguably critically important observation made by much of the previous literature on this topic is that men and women often share very similar motives for joining armed groups. Ideological commitment has long been linked to men's willingness to join armed movements. Yet, as I have noted, political beliefs are also an important driver of women's support for armed groups, including their willingness to transgress social norms by directly participating in violence. In addition, as with male recruits, for female recruits pre-existing political networks represent important conduits through which they enter violent political organizations (Eggert 2018; Kampwirth 2002; Mason 1992; Parkinson 2013; Viterna 2013). Personal exposure to repressive state violence or violence directed against friends or family members also appears to encourage men as well as women to join armed groups (Eggert 2018; Mason 1992; Speckhard 2008; Viterna 2013).[10] Concerns for personal physical security as well as a desire to seek revenge against the perpetrators of such violence are therefore motivations for recruitment shared by both men and women. While few systematic comparisons of men's and women's reasons for participation in rebellion exist, the available evidence suggests that existing theories of mobilization apply to female fighters as well.

Lastly, a handful of recent studies have articulated the possible strategic motives associated with group decisions to recruit and deploy female combatants. Such studies contend that rebel groups rely on female fighters because they are better able to blend into the civilian population and to avoid the scrutiny of security forces, allowing them to conduct more effective and more lethal clandestine operations and terrorist attacks (e.g., Bloom 2011; Cunningham 2003; Hamilton 2007, 108–109). Quantitative analyses find partial support for this claim, showing a strong positive relationship between a group's decision to use terrorist tactics and the presence of female fighters within its ranks (e.g., Thomas and Bond 2015). Other systematic studies provide some evidence of the superior lethality of female suicide bombers (O'Rourke 2009). Consequently, rebel groups may view female combatants as a strategic asset, at least under some circumstances. Despite this claim, to my knowledge, no studies have attempted to systematically examine the influence of female combatants on the groups' likelihood of achieving strategic successes or their ability to accomplish broader political and military objectives.

Women's Agency in Armed Rebellions

In studies of women's roles in armed resistance movements, the agency of the actors represents a normatively and theoretically important question. Agency is a complex concept that is often poorly defined, poorly conceptualized, or completely ignored in studies of armed conflict. Herein, I understand *agency* as the ways in which individuals navigate their social environment, including the various tactics they adopt to survive and potentially thrive in that environment.[11] While I do not explicitly engage the question of women's agency during conflict in this book, it is worth discussing briefly here in order highlight its relevance to the study.

Many female fighters, especially girl soldiers, are victims of abduction or forced recruitment (see Coulter, Persson, and Utas 2008; MacKenzie 2012; McKay 2004). However, coercion is a recruitment tactic employed by many armed groups, and it affects both men and women (see Eck 2014).[12] Similarly, like men, many other women willingly join armed groups as a way to maximize their chances of survival and avoid abuse from government or rebel forces. Still others join rebel movements because doing so provides them with a greater range of opportunities than they enjoyed before the outbreak of the conflict (Coulter, Persson, and Utas 2008; Geisler 2004, 49). Yet, because agency is highly dependent on specific social situations (Utas 2005, 407), the manner in which agency is experienced and the modes of social navigation often differ markedly between men and women. Owing to the different societal expectations and norms imposed upon them, women are often presented with different sets of opportunities and therefore adopt different tactics for survival, resistance, and empowerment during wartime. Women therefore often negotiate their relationships with armed actors and even their positions within armed groups in ways that are distinct from men (Denov and Gervais 2007; O'Gorman 2011; Utas 2005).

Female fighters—even those who are abducted or forcibly recruited—are not simply victims devoid of agency (see Annan et al. 2009; Coulter 2009; McKay 2005; Utas 2005). In developing my arguments about women's participation in armed resistance movements, I therefore assume that (potential) female fighters represent (broadly) strategic actors attempting to navigate a complex and highly uncertain environment. However, I fully recognize that gender norms and expectations influence other actors' (e.g., rebels and civilians) attitudes regarding women's roles during conflict and that these factors often circumscribe or augment the opportunities available to women during wartime (Coulter 2009).

Gender Bias and Abuse in the Ranks

While asserting the nominal agency of female fighters, I nevertheless acknowledge that women participating in armed groups often experience gender bias and mistreatment. Given the hypermasculinity inculcated in the ranks of most organized armed groups, women combatants are often viewed by male troops as outsiders, interlopers, and/or unfit for the rigors of warfare. At minimum, these attitudes routinely result in gender-based discrimination, subordination, and ridicule. In some cases, gender bias and misogyny in the ranks can expose female combatants to severe mistreatment and physical and psychological abuse.

The most systematic manifestation of gender bias within the ranks is the persistence of gender-based divisions of labor within armed groups, even among those organizations that advocate for gender equality. While most armed groups include women in some formal or informal capacity, most of these women serve in noncombat support roles. Comparatively few women participate in combat roles. Moreover, available evidence suggests that only about a quarter of rebel movements active during the post–Cold War period include women in military leadership positions of any sort (Henshaw 2016a). Analyses of women's participation in the Yugoslav Partisan forces suggest that similar gender-based biases in leadership were also present in pre–Cold War armed groups, including those that espoused gender egalitarianism (Jancar 1981, 153–154). In her in-depth study of women in the FMLN, Jocelyn Viterna (2013, 132) observes that while some female members of the organization participated in combat and leadership positions, women were disproportionally represented in support roles in the organization, including cook, radio operator, medic, and so on.[13] Despite ZANLA's public rhetoric of gender equality, mobilization was a gendered process, and sex-based biases and social attitudes contributed to a similar gender-based division of labor in the group (Kesby 1996, 569–570; Lyons 2004, 187–190). Likewise, female members of the Marxist New People's Army (NPA) in the Philippines have complained that they are expected to take on both revolutionary responsibilities as combatants and domestic responsibilities as caregivers (Santos and Santos 2010, 274–275). This "double burden" of serving in combat but being forced to perform "women's work" such as cooking, cleaning, and caregiving was also imposed on the *milicianas* who fought for the anti-Fascist forces in the Spanish Civil War and on female fighters of the Yugoslav partisan forces (Batinic 2015, 148–149; Lines 2009, 170–171, 181–182). These patterns are likely repeated in most other rebel organizations. Consequently, as several scholars have previously noted, peacetime gender hierarchies are often replicated in wartime organizations (Luciak 2001a, 11–13, 18; Reif 1986).[14]

More severe forms of gender-based coercion, mistreatment, and abuse also frequently occur within armed groups (Coulter, Persson, and Utas 2008; Krystalli 2016; McKay 2005; Mazurana et al. 2002). Recent reports suggest that FARC closely controls and monitors women's behaviors, including forcing some women and girls (as young as twelve years old) to use contraception or have abortions (Krystalli 2016; O'Keeffe 2008; Stanski 2006). Women serving in armed groups also commonly face pressures to engage in sexual relationships with men or exchange sexual favors for protection or other benefits. For example, Mary Nash (1993, 280–281) notes that many female fighters in the ranks of Republican and Anarchist forces in the Spanish Civil War complained of being harassed by their male comrades and voiced fears of the difficulties they faced if these sexual advances were refused. Female fighters in the FARC have similarly reported that they were expected to provide sexual services to male combatants (Herrera and Porch 2008, 610). In Liberia and Sierra Leone, many female fighters also served as "bush wives" or "sex slaves" for male commanders during at least some point in their tenure with the rebel movement, often as a means to protect themselves or secure a better standard of living (Coulter, Persson, and Utas 2008, 16; Coulter 2009). Josephine Nhongo-Simbanegavi (2000) likewise notes that despite the group's stated commitment to the advancement of women's rights, female ZANLA recruits often experienced sexual abuse or manipulation at the hands of male guerrillas. Thus, even in armed movements in which women serve as frontline combatants and attain leadership positions, female members and female fighters often become victims of abuse, rape, and sexual exploitation.

While the victimization of female combatants is commonplace during armed conflict, the severity and frequency of abuse appears to vary substantially across groups. Interviews with former female combatants and case studies of women's involvement in some rebel movements often indicate that the women in these movements felt that they enjoyed the respect of their male colleagues and were treated comparatively equally to male combatants. For instance, according to an assessment of gender relations in the FMLN camps written during the final years of the war, women were no longer the servants of men but had become their comrades, and rigid gender barriers had significantly weakened (Carter et al. 1989, 127). Similarly, Viterna (2012, 152–153) asserts that while many scholars have portrayed the rebel camps as places where women were subjected to abuse and harassment, the women in her study sample "resoundingly dismissed" such claims. Rather, they described the guerrilla group as "a family" and insisted that men typically treated female combatants with respect. Angela Veale (2003) similarly reports that sexual abuse in the Tigrayan People's Liberation Front (TPLF) was rare. According to Gisela Geisler (2004, 54–55), female combatants in the South West African People's Liberation Organization (SWAPO) and ZANLA also spoke passionately about the respect male cadre gave them and asserted that in the ranks men and women worked "shoulder to shoulder" and gender was not central to their relationships. Karen Turner (1998, 135) similarly notes that the war in Vietnam eroded gender boundaries in part by highlighting the similar struggles men and women faced: "In the jungles, the tunnels, the clinics, when Vietnamese women turned their gaze toward men, they saw not powerful patriarchal figures but comrades trying to survive with dignity and sometime losing the struggle." The level of mistreatment experienced by female combatants is likely related to the organizational structures and rules of conduct adopted by the rebel leadership. Female former LTTE fighters contend that gender relations in the ranks were generally positive and abuse was rare, partly because the group's leadership imposed strict codes of behavior and punished transgression (Herath 2012, 110–111). Numerous other rebel groups have employed similar efforts to manage gender relationships within the ranks, including prohibitions on sexual or romantic relationships (Taylor 1999, 63–65; Viterna 2013, 154–157).[15]

As this discussion suggests, the experiences of women in armed groups, particularly with respect to their perceptions of equality with male combatants, likely vary both across as well as within groups. Despite some efforts to explore these issues, there has been little effort to systematically explain the factors that influence women's experiences with abuse and mistreatment within armed groups or to explain cross-group variations in the frequency or severity of such abuses. By contrast, a wealth of recent studies has sought to understand the factors that contribute to armed groups' use of sexual violence against civilians. While some studies have asserted the causal role of structural gender inequality in wartime sexual violence (Davies and True 2015), the majority of recent studies have linked variations in such behaviors to group-level organizational and institutional factors. For instance, Dara Kay Cohen (2013a) asserts that rape represents a strategy through which rebel groups that lack internal cohesion and dense internal social ties attempt to forge bonds among combatants. Empirically, groups that rely on abduction or coercive recruitment strategies are more likely to engage in wartime rape.

Other related studies highlight the important role that internal group norms, socialization processes, and the specific institutions created by rebel movements play in determining the frequency with which troops engage in rape and sexual violence (e.g., E. J. Wood 2006, 2009). Amelia Hoover Green (2016) argues that rebel groups that cultivate strong internal cohesion via political education and other institutionalized socialization processes are less likely to engage in sexual violence. Further, she shows that rebel groups with Communist ideologies, which she argues are most likely to possess the features she identifies as causal mechanisms, are comparatively less likely to commit wartime rape than other groups. Taken together, these studies suggest leaders can exert substantial control over the level of sexual violence their troops commit. However, their ability to control such violence is often dependent on the investments they made early in conflict to developing shared norms, command hierarchies, and disciplinary mechanisms. Notably, this research also demonstrates the important ways in which group ideology can shape rebel behavior and group attitudes toward women. I return to this important point in chapter 1.

Gender Norms and Armed Conflict

The arguments that I put forth in subsequent chapters are largely predicated on the observation that societal gender norms exert a substantial influence on women's roles during periods of armed conflict. While I have briefly alluded to the role of societal gender norms and gender-based divisions in determining the presence of female combatants in rebel movements, it is useful to more fully elaborate on this point before I proceed. This discussion emphasizes the historical and cross-cultural association of men (and masculinity) with violence and warfare and women (and femininity) with innocence, virtue, and peace. It also highlights the manner in which war often reifies gender stereotypes and reinforces gender-based divisions of labor. During periods of violent armed conflict, men therefore face strong social pressures to conform to traditional masculine roles and participate in combat while women are discouraged from undertaking such activities and are instead expected to perform more traditionally feminine roles such as providing care, support, and encouragement to male fighters. These social expectations have implications for rebel leaders' attitudes regarding the recruitment of women and their deployment in combat. The countervailing influences of these gender norms/biases and wartime resource demands faced by rebel leaders represent the core tension I address in chapter 1.

Though exceptions can be easily found, across diverse cultures and time periods the vast majority of warriors have been male.[16] Joshua Goldstein (2001, 10), for instance, asserts that women likely represent less than 1 percent of warriors throughout recorded history. However, the frequency of these exceptions varies across different cultures and historical periods. In recent years, the frequency of women's participation in national armed forces has increased, particularly in Western states. As of 2017, a dozen countries (representing roughly 6 percent of all nation-states) permitted women to serve in frontline combat positions in their national military forces (Fisher 2013; Keating 2012).[17] Yet in all cases, female combatants represent only a very small fraction of a state's total combat forces. Available estimates suggest that women constitute somewhere between 2 and 7 percent of combat troops in the military forces of those states that allow them to serve (see Alexander and Stewart 2015; Ben-David 2017; Mulrine 2013; Phippen 2016). Moreover, while rebel movements are comparatively more likely to deploy women in combat (see J. Goldstein 2001, 77–79), the prevalence of female combatants in the ranks of these groups is still quite low in most cases. The majority of rebel groups exclude women from combat, and in only about one in five rebel groups does the number of female fighters exceed 5 percent of the group's total combat forces (Wood and Thomas 2017). Despite the important contributions women have made (and continue to make) in many armed groups, the battlefield represents a heavily gendered, intensely masculinized environment.

Explanations for the overall scarcity of female combatants vary and include essentialist arguments regarding innate sex-based preferences over the use of violence, physical and cognitive differences between men and women, and the influence of socialization and cultural expectations. A few feminist scholars, for example, have asserted that women are inherently more opposed to war and resistant to militarism compared to men. Virginia Woolf ([1938] 1963, 6) appeared to adopt this perspective when she wrote that for men there is "some glory, some necessity, some satisfaction in fighting that [women] have never felt or enjoyed." Sara Ruddick (1989), writing more than half a century later, embraced a similar view when she asserted that women are inherently more pacifistic than men and that this (comparatively) more peaceful nature largely stems from their biological and social roles as mothers. However, these essentialist arguments generally fail to explain why many women are willing participate in armed groups when provided the opportunity to do so.

Differences in physical capabilities offer a partial explanation for the dearth of female combatants. Joshua Goldstein (2001), for instance, finds some evidence that sex-based differences in average size and strength may exert a significant deterrent effect on women's participation in military forces. More systematic and focused studies of women's performance in combat actual units also acknowledge important physical differences between men and women in terms of strength and stamina (Epstein et al. 2013; Finestone et al. 2014), and some studies suggest that all-male combat units generally perform better than mixed-gender units (Peralta 2015).[18] However, other analyses conclude that at least a subset of the female population is capable of meeting the physical standards militaries establish for combat units and can successfully perform combat duties. They further suggest that the integration of women into combat units is logistically feasible and does not substantially degrade unit morale or performance (e.g., Cawkill et al. 2009; Epstein et al. 2013; Finestone et al. 2014; UK Ministry of Defense 2014). The recent graduation of two women from the U.S. Army's grueling Ranger School further suggests that at least some women can meet even the most rigorous physical and mental standards expected of soldiers (Lamothe 2016). These findings therefore undercut the explanatory power of the capabilities difference argument. While women are smaller and less physically strong on average, reducing the number of women physically capable of enduring the rigors of armed combat, most populations will include a substantial number of women who would otherwise appear physically qualified to become frontline combatants.

Sociocultural factors likely play the most significant role in the gendering of warfare. Joshua Goldstein's (2001) extensive investigation of the relevant literature leads him to conclude that embedded gender norms within most societies orient men toward war and conflict and women toward maternal roles. Moreover, women play a critical role in the process of cultivating male warriors by providing support, encouragement, and rewards to men who conform to these gendered expectations and castigating, shunning, and shaming men who deviate from them. Due to their powerful influence on the moral, social, and psychological development of their children, women play a key role in raising boys and girls that fulfill their expected roles. Ultimately, he asserts that the "cultural molding of tough, brave men, who feminize their enemies to encode domination" represents the most compelling explanation for the near-universal gendering of war roles (2001, 406). The assertion that the exclusion of women from combat results from sociocultural factors rather than innate physical or psychological differences between men and women largely dovetails with longstanding feminist critiques of gender and warfare. Principally, the general absence of women from the battlefield—as well as frequent attempts to erase or minimize the contributions of the relatively few female fighters that might have participated—is a reflection of deeply embedded social and cultural norms regarding the "appropriate" division of labor between genders. As Jacklyn Cock (1991, 132) asserts, divisions of labor along gender lines in state militaries reflect the cultural norms of the society and current ideas of the proper roles for men and women in the society.

These attitudes appear to intensify during war. Despite the varying rigidity of gender norms, differing perspectives on women's rights, and wide discrepancies in women's access to educational and occupational opportunities across societies, war often evokes highly gendered attitudes and reifies traditional gender roles. As Jean Elshtain (1987, 4) asserts, "in time of war, real men and women . . . take on, in cultural memory and narrative, the personas of Just Warriors and Beautiful Souls. Man is construed as violent, whether eagerly and inevitably or reluctantly and tragically, woman nonviolent, offering succor and compassion." These themes and images are contrived, and they do not necessarily reflect the reality of women's and men's identities or actions during war. Yet, they serve—intentionally or inadvertently—to reaffirm the female position of innocent (and often virtuous) noncombatants and the male role of valiant protectors and warriors. Furthermore, in this narrative women serve not only as innocent potential victims of war in need of protection by (and ironically from) men but also as the reward offered to male protectors for dutifully risking their lives in war (Elshtain 1987). Drawing on this earlier work, Laura Sjoberg (2010, 55) therefore contends that women are thus at once the "object of fighting and the just purpose for war." In this sense, this imagined essentialist distinction between violent (but valiant) men and innocent, pacifist women has the effect of justifying the act of war itself.

Within the Just Warrior–Beautiful Soul narrative, men and women are expected to play opposite but mutually reinforcing roles, which ultimately function to reaffirm the traditional gender hierarchies and divisions of labor.[19] Moreover, the performance of these highly gendered roles is intended to facilitate and sustain the mobilization of the population during war. Specifically, social pressures to affirm their masculinity encourage men to risk their lives and health in defense of the country; parallel social norms and expectations press women to encourage and reward men who accept their masculine obligation and to shame men who seek to evade those obligations (De Pauw 1998, 19; J. Goldstein 2001, 272–274; Sjoberg 2014, 32–33). Interestingly, while this narrative normally assumes that femininity is an inherent trait of women, resulting in women being more pacifistic, compassionate, and nurturing, men must overtly demonstrate their masculinity and earn their manhood through concrete actions, such as by participating in armed combat.[20]

Political organizations and national governments have historically been acutely aware of the power such gender tropes have in shaping prospective (male) recruits' attitudes toward participating in armed conflict on behalf of the state. Wartime recruitment propaganda illustrates the ways in which state militaries and governments manipulate the gender narrative summarized above in order to encourage men to fight threats at home or enemies in foreign lands (Rupp 1978; Shover 1975; Sims 2000). For example, in England during the World War I and briefly during World War II, groups of women handed out white feathers to British men who had not enlisted in the military. The intention and symbolism of these actions by the women of the "White Feather Brigades" was clear: to remind men of those whom they fought to defend, highlight what rewards might await the brave men who enlisted, and to label as cowards—and thus ineligible for such rewards—those men who evaded their masculine obligations (Gullace 1997, 182–193).[21] Similarly, during the Spanish Civil War, Republican activists and politicians urged women to publically denounce and shame men who ignored the call to arms (Nash 1993, 271). The mere presence of female combatants has also served as way to shame reluctant men into volunteering for combat. For instance, the Czarist government of Russia formed a female battalion during World War I largely for this purpose, a tactic that appears to have succeeded (J. Goldstein 2001, 73–74). Anecdotal evidence from a variety of cases, including civil wars in El Salvador, Ethiopia, Turkey, and Vietnam, likewise illustrates that female guerrillas have often proven effective at encouraging, cajoling, or shaming men into joining rebel movements (Donnell 1967; Marcus 2007; Tareke 2009; Viterna 2013).

From this discussion, I draw several assumptions on which I base the arguments that follow in this book. First, historically and across disparate cultures, war is viewed as an overtly masculine pursuit (e.g., J. Goldstein 2001). Second, despite the recognition that women can perform admirably in combat, deeply embedded gender norms in most societies position men to assume the role of fighters and defenders and women to act as caregivers and victims. Moreover, warfare serves to reify preexisting gender beliefs, hierarchies, and norms, pushing men and women to assume their traditional gender roles. Third, in most societies, the presence of female combatants represents a dramatic departure from existing gender norms. Because of this, the decision to recruit women and deploy them in combat is not one that rebel leadership undertakes lightly. Specifically, both the gender biases of a group's leadership and the possibility of alienating relevant constituencies influence this decision. Understanding how these factors shape rebel leaders' (and audiences') perspectives on women's participation in political violence helps to highlight the relevant constraints and incentives that shape leaders' decisions to recruit female fighters or to exclude women from combat.

Plan of the Book

Despite being underrepresented in most rebel movements and, in many cases, rendered invisible in the aftermath of the conflict by both the international community and by the movements for which they fought (see, e.g., McKay 2005; Mazurana et al. 2002), female combatants have been present on the frontlines of many contemporary armed conflicts. In the following chapters of this book I seek to explain why some armed resistance movements actively recruit women and deploy them in combat while others exclude female combatants, and to better understand what impact this decision ultimately has on the groups that make it. Having provided a brief introduction to the central arguments as well as a discussion of the scope of conditions and some core assumptions, I now provide a roadmap to the remaining chapters.

In chapter 1, I develop a set of arguments to explain why some rebel movements have chosen to incorporate women into their combat forces—sometimes in substantial numbers—while others have instead excluded them. While acknowledging the importance of understanding why women choose to join rebellions, I focus my attention in this chapter primarily on explaining the conditions under which rebel leaders become willing to ignore traditional gender norms (and their own gender biases) in order to create opportunities for female supporters to assume combat roles within the armed movement. I argue that a combination of strategic and ideological factors shapes a rebel leader's decision to incorporate women into the group's fighting forces and deploy women in combat: (1) the severity of the conflict and the corresponding resource commitment necessary for the rebel organization to maintain its war against the incumbent regime, (2) the group's ideology, and, finally, (3) the interaction of these factors. I focus particularly on the way that group ideology conditions the relationship between resource demands and female recruitment.

I consider the potential strategic benefits associated with the decision to recruit female fighters in chapter 2. The arguments therein represent a logical extension of those proposed in chapter 1. I argue that female fighters potentially benefit the groups that employ them in multiple ways. Consistent with the logic outlined in chapter 1, the leadership's decision to recruit women directly expands the supply of available troops, thus easing human resource constraints that might otherwise adversely impact the group's ability to achieve its goals. In addition, I discuss the ways in which rebel groups utilize female combatants in their propaganda efforts and the potential role that female combatants can play in assisting the group in securing support from external actors.

Chapter 3 probes the validity of the arguments put forth in chapters 1 and 2 by examining the evolutions of women's participation and the implications of female combatants in three diverse rebellions: the PKK and its allied militias in Turkey and Syria, the LTTE in Sri Lanka, and ZANLA and ZIPRA in Rhodesia/Zimbabwe. Each case broadly corresponds to each of the arguments made in the theory chapters. Moreover, each clearly illustrates one or more of the mechanisms posed in the theory. For instance, the PKK case study helps illustrate the role of ideology in shaping leaders' attitudes toward female combatants (and how these views evolve over time). Yet, as the other two cases demonstrate, Marxist ideology is not a necessary condition for the inclusion of female combatants. Each also highlights the importance of increasing conflict intensity and tightening resource constraints in motivating rebel leaders to incorporate women into the groups' fighting forces. All three of the cases demonstrate how rebel groups utilize the presence of female fighters in their ranks to solicit support from external actors, including diaspora communities and transnational activist networks.

Chapters 4 and 5 present results from the quantitative analyses used to assess the hypotheses presented in the earlier chapters. In chapter 4, I introduce and provides a brief but detailed discussion of the dataset used to assess women's participation in rebellions as well as a the quantitative tests of the hypotheses based on the arguments presented in chapter 1. The analyses provided in this chapter suggest that rebel groups engaged in conflicts that result in greater numbers of combatant deaths and causalities are more likely to recruit female combatants and more likely to include a larger proportion of female combatants than less severe conflicts. They also imply that the ideology a rebel group adopts influences the presence and prevalence of female combatants. Moreover, there is evidence that group ideology conditions the more general relationship between resource demands and female recruitment.

In chapter 5, I subject the hypotheses on the implication of female combatants presented in chapter 2 to rigorous empirical scrutiny. The results of the quantitative analyses presented in that chapter suggest that rebel organizations with a greater prevalence of female combatants field larger combat forces than their counterparts that exclude female combatants or recruit in very small numbers. This chapter also describes and presents the results from a novel survey experiment designed to assess the effects of female combatants on audience perceptions of a rebel movement. Results from this experiment suggest that the presence of female combatants in an armed group improve respondents' overall support for the group's goals. Finally, I present the results of analyses examining the relationship between female combatants and external support for rebel movements, particularly by transnational advocacy networks and diaspora communities. These analyses provide preliminary evidence that the presence of female combatants can influence the likelihood that a group receives external support.

The final chapter summarizes the central arguments and findings, considers their implications, and discusses their relevance to understanding the contemporary rebellions. It highlights the important—yet often underappreciated role—female fighters play in these conflicts and suggests how a better understanding of women's participation in armed groups might contribute to additional theoretical insights into the behaviors in which armed groups engage and the strategies and tactics they employ.

CHAPTER 1

Why Rebels Mobilize Women for War

Between 1942 and 1945, thousands of women joined the National Liberation Army (NLA) to fight against the Nazi occupation of Yugoslavia. Female fighters ultimately made up as much as 15 percent of the Partisan fighting forces, with many serving in frontline combat positions (Batinic 2015, 131–132; Jancar 1981). Despite the scope and scale of their eventual participation in the group, women were largely absent from the Partisan combat forces at the beginning of the conflict. While a few women served in the NLA at the insurgency's outset, they were deployed almost exclusively in support roles (e.g., nursing, logistics, administration, etc.) and were generally prevented from directly participating in combat. By early 1942 this practice changed, and the NLA began training and arming female members and allowed some female recruits to serve in combat positions (Batinic 2015, 127–129). A series of recruitment campaigns undertaken by the NLA later that year further expanded the number of women in its ranks and contributed to the relatively high proportion of female fighters that participated in the resistance movement throughout the remainder of the conflict (130).

To explain the observed change in the NLA leadership's position on the role of women in the group's fighting force, previous studies have pointed to a combination of ideological and strategic factors. First, the NLA was closely connected to and largely traced it origins to the Communist Party of Yugoslavia, an organization with a long history of advancing women's rights and encouraging women's political activism (Jancar 1981). As such, the incorporation of women into the Partisan forces was largely consistent with the prevailing Marxist ideology of the organization. In addition, some scholars contend that Partisan leaders sought to use the presence of female combatants to stress the gender egalitarian values of the movement and thus showcase their commitment to Marxist ideology to powerful Communist allies such the Soviet Union (Batinic 2015, 128–129; Jancar 1981, 150). These factors help explain the willingness of Tito and other NLA leaders to recruit female combatants. However, the acute resource demands the group experienced during winter of 1941–1942 arguably provide a more compelling explanation for the timing of the decision and the scope of the recruitment effort (Batinic 2015, 129–130). During this window of time, Partisan forces suffered heavy losses as they simultaneously attempted to repel multiple German offensives and engaged in frequent clashes with Chetnik forces. Manpower shortages resulting from these events left the Partisan forces weakened and placed the NLA's long-term survival in jeopardy. The leadership's decision to recruit women and deploy substantial numbers of these recruits in combat therefore corresponds to a period of acute resource constraints.

As I argue in this chapter, far from simply explaining the timing and scope of women's participation in this specific case, the joint roles of ideology and resource constraints in shaping rebel leaders' decision to mobilize women in war are broadly generalizable to a large number of civil conflicts. Using insights from existing "demand-side" models of rebel recruitment, I link female recruitment to the combination of the severity of the resource constraints imposed on the group as well as to leaders' perceptions of the relative costs associated with opening recruitment opportunities to women as a means of addressing those constraints. More specifically, I argue that sharp increases in resource demands—such as those that occur during periods of intense fighting or following substantial troop losses—incentivize rebel leaders to recruit female fighters as a strategy of preserving or expanding the rebellion. However, the gender beliefs and attitudes of a group's leadership as well as those of the constituencies on which the group relies for support determine the costs the leadership expects to incur from recruiting women and deploying them on the battlefield. The political ideology embraced by a rebel group—which I contend conveys important information about those expected costs—conditions the influence of resource demands on leaders' decisions to employ female combatants.

Consequently, despite their initial reluctance, secular rebel groups have become increasingly willing to utilize female combatants as war-related human resource constraints intensify. By contrast, armed groups that embrace ideologies oriented toward reinforcing or reaffirming traditional social hierarchies, which are comparatively more sensitive to the costs associated with recruiting female combatants, remain unlikely to recruit women for war even as sharply rising resource demands potentially jeopardize the movement's ability to achieve its goals.

The Political Economy of Rebel Recruitment

The likelihood that a rebel movement survives and ultimately achieves its political and military objectives heavily depends on its ability to mobilize human and material resources (e.g., Cunningham, Gleditsch, and Salehyan 2009; Hultquist 2013). As such, resource mobilization, particularly the ability to recruit and retain collaborators and combatants, represents one of the central tasks of any successful rebel organization (Gates 2002; Leites and Wolf 1970, 32–34). To achieve this objective, rebels select from among a variety of recruitment strategies. Rebel leaders determine a set of incentives to offer to potential recruits, which typically includes some combination of coercive threats and promises of material (or nonmaterial) benefits. They likewise select the geographic areas in which they choose to recruit, the populations they intend to target for recruitment, and the types of recruits they are willing to accept.

In order to explain the recruitment strategies that rebels adopt—as well as the relative success of these strategies—previous studies have often relied on insights from industrial organization theory and labor economics (e.g., Gates 2002, 2017; Humphreys and Weinstein 2008; Weinstein 2007). While the specific theoretical models employed by these studies vary, most adopt some form of basic supply-demand framework. According to such approaches, rebel recruitment is a function of both the supply of individuals available to fight for the rebel group and the rebel leadership's demand for these potential recruits. Furthermore, this framework implies that the ratio of the supply of potential recruits to the group's demand for them determines the specific recruitment tactics it adopts. For example, scholars have previously employed this framework to explain the prevalence of child soldiers in armed groups (Achvarina 2010; Achvarina and Reich 2006; Andvig and Gates 2010; Beber and Blattman 2013). According to these studies, factors that increase the supply of children available for recruitment, such as rebel access to refugee camps and the number of displaced and orphaned children among the local population, increase the likelihood that rebels will recruit children for combat. Additionally, rising demand for child soldiers, which is often related to an insufficient supply of adult recruits, increases the prevalence of child soldiers. In a similar manner, scholars have also linked the balance of persuasion and coercion rebel movements employ to mobilize support and acquire recruits with variations in the intensity of the human resource demands they face (Eck 2014; Gates 2002; Leites and Wolfe 1970). Particularly, these studies suggest that rebels increasingly replace voluntary appeals for support with more coercive tactics as the demand for new recruits outstrips their ready supply.

Scholars have also recently begun employing aspects of this framework to better understand women's mobilization for armed conflict. In the context of female recruitment, supply represents the available population of women who would willingly respond to the group's offer of membership, while demand reflects the rebel leadership's willingness or desire to create opportunities for women to participate in the armed movement as combatants. Broadly speaking, most research investigating the conditions that encourage women to join an armed group directly or indirectly addresses the supply-side factors associated with female recruitment. Such studies have, for example, found that women's exposure to state-sponsored violence (especially sexual violence) and their subsequent desire for security or revenge (Bloom 2011; Speckhard 2008; Viterna 2013); the role of prevailing gender norms and women's prewar participation in social, economic, and political processes (Reif 1986; Thomas and Wood 2018); women's previous involvement in activist networks (Eggert 2018, 11; Kampwirth 2001; Mason 1992; Viterna 2006); and the appeal of revolutionary or gender-inclusive political ideologies (Kampwirth 2001; Molyneux 1985) serve as important supply-side factors explaining women's participation in armed groups.

These analyses have produced valuable insights about the dynamics of female recruitment. However, as Thomas and Bond (2015, 488–489) contend, existing scholarship on female fighters has disproportionally focused on supply-side factors and has often ignored or downplayed the central role that demand-side factors play in determining the presence of women's participation in rebel and terrorist organizations. In their effort to examine the role of demand-side factors, Thomas and Bond therefore focus on understanding organizational preferences regarding gender diversity in the ranks and seek to identify the group-level characteristics that encourage or discourage the recruitment of women and their deployment in combat. Ultimately, they conclude that the perceived tactical advantages of using female combatants (e.g., in terror attacks), whether the group embraces a gender-inclusive ideology, and the leadership's assessment of the potential threat female recruits pose to group cohesion and public support influence the group's demand for female combatants. Moreover, they assert that such demand-side factors have more explanatory power than many common supply-side factors.

Following these insights, my theory privileges the role of demand-side factors in explaining the mobilization of women for war. I focus primarily on demand-side factors for two specific reasons. First, as I discussed in the introductory chapter, anecdotal evidence suggests that even in societies characterized by strict gender-based divisions of labor and the presence of rigid gender norms, a subset of the female population is willing to take up arms on behalf of an armed movement provided they are given the opportunity. This appears to be the case even for groups whose beliefs contradict liberal notions of women's rights and gender equality. For instance, substantial numbers of women have volunteered to fight on behalf of radical Islamist groups such as the Palestinian Islamic Jihad (PIJ), Hezbollah, Hamas, Al-Shabaab, the Tehrik-i-Taliban, and (recently) ISIS. Substantial numbers of women have also fought on behalf of a wide variety of other violent movements that made no explicit appeals to gender equality or women's rights, including the PIRA in Northern Ireland, the Unidad Revolucionaria Nacional Guatemalteca (URNG), Chechen insurgents in Russia, and the National Resistance Army (NRA) in Uganda. That women were willing to mobilize in support of these groups, and in some cases risk their lives in combat, implies that locating a supply of potential female recruits seldom represents an acute problem for rebel movements.

Second, the availability of opportunities for women to (formally) participate in organized political violence is inextricably linked to rebel leaders' willingness to create those opportunities (Thomas and Bond 2015, 489). Regardless of the willingness of a subset of the population to participate in armed rebellion, the organization's leadership maintains ultimate authority over the recruitment strategy it pursues, including how it defines its pool of potential recruits and whom it ultimately accepts into the movement (see Weinstein 2007). The leadership likewise determines the specific tasks recruits undertake, the roles they occupy in the movement, and their ability to advance in the ranks. As I discussed previously, in many (if not most) rebel organizations, women are either confined to noncombat support roles or completely absent. This implies that rebel leaders often eschew female recruits or restrict their participation to noncombat roles even when some subset of the female population is willing to fight and die on their behalf.

Acknowledging the leadership's authority over the group's recruitment decisions minimizes but does not eliminate the influence of supply-side factors in determining the prevalence of female combatants.[1] Nor does prioritizing leader strategy formation imply that women are denied agency in the recruitment process. Where recruitment is a nominally voluntary process, I view female potential recruits as agentic actors whose willingness to participate in armed rebellion is shaped by their individual beliefs, preferences, and experiences. Thus, as with male recruits, factors such as prewar political engagement, war-related grievances, and patterns of interaction with armed actors influence their decision to participate (Viterna 2006, 2013). Moreover, where substantial numbers of women have successfully gained entry to armed groups, they have sometimes successfully influenced the group's strategies and goals. For example, Jennifer Eggert (2018) contends that the presence of female fighters in many of the armed groups in the Lebanese Civil War in the 1970s and 1980s was a direct result of women's efforts to lobby rebel leaders to include them. Anecdotal evidence from numerous other rebellions, including the Sandinista Revolution, Mozambican Independence War, Zimbabwean War of Liberation, and Sri Lankan Civil War, similarly demonstrates that once admitted to the movement women have persuaded reluctant rebel leaders to expand women's roles and responsibilities, including allowing them participate in combat or assume leadership roles (C. Johnson 1992, 160–161; Luciak 2001a; Lyons 2004, 109–110; Stack-O'Connor 2007b, 97–98).

Similarly, while rebel leaders routinely subordinate women's interests and issues, the extent to which the group relies on female combatants may determine whether or not the advancement of women's rights becomes a central part of the rebel group's political platform. For example, as female participation in the PKK increased in the early 1990s, the group's attention to issues of women's rights and their oppressed status in the Kurdish community increases as well (see van Bruinessen 2001, 105–106; White 2015, 146–149). As these examples illustrate, even though the leadership retains ultimate discretion over women's entry into the movement and the roles they play, women are sometimes able to successfully advance their collective interests within the organization, particularly when the organization relies heavily on their labor and their support.

The key benefit of the framework discussed above is that it permits me to theorize directly about rebel strategy formation. I focus explicitly on the decision-making processes of rebel leaders in order to identify the conditions under which rebel leaders are more likely to perceive that the benefits to recruiting, arming, and deploying women in combat roles outweigh the attendant costs of their inclusion in the armed wing of the unit. I view the recruitment of female combatants as a strategic decision; moreover, it represents only one of multiple potential resource mobilization strategies available to rebel leaders. Whether or not rebel leaders ultimately adopt this strategy depends on the intensity of the demand for such troops and the potential costs they expect to incur from adopting this strategy.

Resource Demands and Female Recruitment

While firms in competitive markets typically respond to increases in labor demands by offering higher wages, the asymmetric nature of internal armed conflict as well as the high risks associated with joining an armed rebellion often inhibit rebel ability to offer sufficient material incentives to address conditions of excess demand. Because the inability to address human resource shortages jeopardizes group survival and diminishes the odds of success, rebel leaders often adopt alternative strategies to address recruitment shortfalls. When conditions of labor scarcity prevail, rebels become more likely to adopt violence and forcible recruitment strategies to acquire necessary resource inputs (see Andvig and Gates 2010; Eck 2014; R. Wood 2014). Violent mobilization strategies, including gunpoint recruitment, abductions, and press-ganging, are indeed a common feature of contemporary internal conflicts (see Cohen 2013a). However, they do not represent the only strategies through which rebels seek to remedy human resource shortfalls; nor are they necessarily the preferred strategies of most groups. Locating alternatives to forcible recruitment may be particularly desirable given that overreliance on such strategies invites international condemnation and costly sanctions that might ultimately impede the group's ability to achieve its political and military objectives (see Jo 2015).[2]

An alternative strategy available to many rebel organizations involves redefining the pool of potential recruits and extending opportunities to new groups or subpopulations that the leadership had previously overlooked or intentionally excluded from participation. While rebel leaders might appear to have only minimal ability to manipulate the supply of potential recruits, the historical record demonstrates that state militaries and rebel organizations have often sought to redraw the boundaries separating acceptable and less acceptable recruits. These efforts typically become more common as conflicts drag on and resource demands increase. For example, during World War I the U.S. government initially required all male citizens between the ages of twenty-one and thirty-one to register for the draft but later expanded this requirement to the ages of eighteen to forty-five.[3] Similarly, acute manpower shortages during the brutal Iran-Iraq War led the Iranian government to call both adolescent boys and elderly men (though mostly the former) into military service (Singer 2006, 22). Military leaders typically exclude children and adolescents from their primary recruitment pool, both because they are widely perceived as suboptimal combatants compared to adults and because deploying them violates domestic and international norms and therefore might provoke condemnation and retaliation by the international community.[4] Nonetheless, the voluntary and involuntary recruitment of adolescents and children by state and rebel forces becomes more common when sufficient numbers of adult recruits are unavailable (Achvarina 2010; Lasley and Thyne 2015; Twum-Danso 2003). Related options for expanding the pool of potential recruits include enlisting foreign fighters (Malet 2013) or extending recruitment opportunities to members of ethnic communities other than those that form the movement's core constituency (Kalyvas 2008).[5]

How and to what extent rebel leaders seek to define and redefine the pool of potential recruits varies substantially across groups and conflicts. However, factors such as the total size of the subpopulation initially excluded from the supply of acceptable recruits, the physical capabilities of the persons within them, and the anticipated costs associated with recruiting individuals from these subpopulations all likely influence rebel leaders' decisions regarding who qualifies as an acceptable or unacceptable recruit. All else being equal, rising resource demands should offset some of the concerns that initially led leaders to exclude specific subsets of the population from consideration for recruitment. Thus, leader willingness to accept recruits that they previously viewed potentially costly increases as resource demands become more acute.

Assuming that at least some subset of the female population possesses the physical and psychological qualities necessary to serve in combat and that leaders anticipate no substantial additional costs arising from their recruitment, extending recruitment opportunities to this subset of the population permits leaders to relax the constraints on their supply of potential recruits.[6] Arriving at this conclusion does not require assuming that rebel leaders view female recruits as perfect substitutes for male recruits. Even if rebel leaders perceive female combatants as comparatively less desirable than male troops, resource-maximizing leaders should still elect to recruit female fighters because the value of even suboptimal combatants typically exceeds the costs of severe recruitment shortfalls. Consequently, rising resource demands should increase the likelihood that rebel leaders recruit women and deploy them in combat.

The discussion above conceptualizes the demand for female combatants strictly in terms of the human resource needs of the organizations that employ them. Yet, it is important to note that other forms of demand that are not strictly defined by resource needs may also incentivize rebel leaders to recruit, train, and arm women. For example, previous studies suggest that rebel groups often view women as comparatively more effective in carrying out specific types of attacks, such as sabotage, assassinations, terrorism, and other forms of covert violence (e.g., Cunningham 2003; Hamilton 2007, 108–109; O'Rourke 2009). In this sense, women's perceived superior ability to blend in, evade capture, and get close to targets is viewed as a specific skill that is in high demand for groups that engage in such tactics. The desire to recruit women should therefore be relatively high among groups that routinely utilize such clandestine activities (see Thomas and Bond 2015).

Demand may also increase as rebel leaders begin to recognize women's potential to deliver additional strategic benefits to the group and seek to capitalize on it. Many rebel groups have instrumentalized women's presence in the movement in an effort to garner attention, solicit external support, and encourage recruitment (see Cunningham 2003; Speckhard 2008; Viterna 2013). By including images of female combatants in their propaganda materials, explicitly employing women as recruitment agents, or facilitating international media access to its female cadres, rebel leaders may seek to secure strategic benefits from high rates of female participation that are not explicitly connected to the overall human resource needs of the organization. However, these benefits may only become apparent after a substantial number of women have already gained entry to the movement (Stack-O'Connor 2007b, 97). To the extent this is the case, it is difficult to make the case that the expectation of these benefits represents a central motivation for rebel leaders' initial decision to recruit female combatants. I therefore reserve the discussion of these downstream benefits for a subsequent chapter and focus explicitly on the role of human resource constraints in determining when and why rebel leaders recruit female combatants in this chapter.

The Dilemma of Female Fighters

A potential limitation of the strictly demand-focused recruitment model discussed above is that its predictions rest on the assumption that rebel leaders perceive no substantial costs associated with the decision to recruit female combatants other than the (potential) marginal physical inferiority of these recruits. If this assumption were realistic, we would expect to see rebel movements routinely training, arming, and deploying women in combat whenever they experience human resource demands that dramatically exceed the available supply. Moreover, given that rebel groups often struggle to expand the movement beyond an initial cadre of core supporters, we might expect to see women routinely enter armed resistance movements early in the conflict, when groups are still extremely weak relative to the governments they seek to challenge.

Yet, this is often not the case. Rather, the majority of rebel movements—even those facing chronic manpower shortfalls—refrain from deploying women in combat. Moreover, those groups that eventually elect to extend recruitment opportunities tend to do so relatively late in the conflict or only once the group is well established (Thomas and Bond 2015; Wood and Thomas 2017). Indeed, rebel leaders are often initially reluctant to incorporate women into their group's fighting force. For instance, while female combatants eventually constituted as much as a third of the Sandinista fighting forces, traditional views of women's roles in Nicaraguan society initially constrained their participation, and the organization refrained from recruiting women during the early years of the rebellion (Chuchryk

1991, 143–144; Luciak 2001b, 195). Similarly, female combatants were largely absent from the Frente de Libertação de Moçambique (FRELIMO) in its early years; however, by 1967—three years after the onset of major conflict activities—the group's leadership created a women's combat detachment (Urdang 1984, 165; Geisler 2004, 50).[7] Lastly, while women were well represented in the African National Congress (ANC) as a whole, through the early 1970s uMkhonto we Sizwe (MK), the organization's armed wing, included very few female fighters. However, by the early 1980s women constituted roughly 20 percent of MK's fighting force (Cock 1991, 41, 161–162; Geisler 2004, 51). In addition to demonstrating leaders' reluctance to recruit women, these examples also underscore the assertion that many women are willing to participate in a movement, provided an opportunity exists.

An important implication of this discussion is that acute conflict-induced resource demands alone—which rebel groups routinely experience during their wars against the state—cannot fully explain the decision to employ women in combat roles. A compelling explanation for variations in the prevalence of female combatants across armed groups should therefore also explain why so many groups resist recruiting, arming, and deploying female fighters, even as the demand for human resource inputs sharply increases. Addressing this question requires accounting for the factors that influence the strength of rebel leaders' biases against deploying women in combat roles and their perceptions of the potential costs associated with this strategy. I therefore consider how both the gender attitudes of rebel leaders as well as the attitudes and beliefs of key segments of the society in which the conflict occurs shape leaders' decisions regarding how to respond to rising resource demands and their overall willingness to use female combatants as a strategy for addressing these demands.

The belief that war is a definitively masculine enterprise is deeply entrenched in most societies. While men are rewarded for risking their lives in combat, women are often characterized as potential victims in need of (male) protection as well as the rewards provided to male combatants for their sacrifices (e.g., Elshtain 1987; Sjoberg 2010). One of the consequences of this dynamic is that women are routinely discouraged from participating in armed groups (J. Goldstein 2001). Military culture is universally hypermasculinized and characterized by beliefs in stereotypical gender roles and the devaluation of feminine traits (Cock 1991, 59–60; Baaz and Stern 2009; Wood and Toppelberg 2017, 624). Military leaders, including those of rebel movements, typically view female combatants as incompatible with their beliefs about *how* wars should be fought and *who* should fight them. Particularly, leaders often resist incorporating women because of concerns about their impact on group cohesiveness and combat preparedness (J. Goldstein 2001, 96–98, 200–201; Woodward and Winter 2004). Not only do they doubt women's ability to endure the physical and emotional rigors of combat, they often fear that the inclusion of female combatants will erode morale, create undue distractions for male soldiers, and generate discord in the ranks.

While military leaders often harbor concerns about these issues, the depth of resistance to female combatants appears to be context dependent. For example, where rebel leaders have had previous exposure to women engaging in high-risk collective or political activism, they appear to be more amenable to incorporating them into an armed movement (Thomas and Wood 2017). Moreover, the specific gender ideologies of the leadership are expected to have a direct impact on the group's recruitment of female combatants. Specifically, if the leadership is committed to gender equality and women's rights in society, the group is more likely to recruit female combatants (Thomas and Bond 2015).

The attitudes and beliefs of the constituencies on which rebels rely for support likewise influence the leadership's willingness to utilize female combatants. Rebel movements often depend heavily on local communities and networks of local power holders (e.g., religious leaders, tribal councils, clan leaders, village headmen, etc.) for support. Moreover, their behaviors and organizational structures are strongly influenced by the social bases from which they arise (Staniland 2014). Directly challenging or contravening the norms and values embraced by these constituencies can prove costly for armed movements, particularly when those values are deeply entrenched and highly salient. Behaviors that radically diverge from local norms or that challenge local customs are likely to create tensions between the rebel group and these communities. Successful rebel movements must therefore carefully navigate their relationships with local constituencies. Positioning women in roles for which they are trained, armed, and instructed to commit violence on behalf of a political movement represents a dramatic deviation from the historical norm and may be perceived by many observers as a direct challenge to deeply engrained social norms regarding women's place in society. Where women's participation in organized violence defies prevailing gender norms and threatens existing social hierarchies, rebels potentially risk alienating the local populace by utilizing female fighters. Consequently, the gender beliefs and attitudes of the rebel group's constituent community partially determine the costs the group expects to incur from using female recruitment to ameliorate human resource constraints.

Gerakan Aceh Merdeka's (GAM) secessionist rebellion illustrates how expectations of these costs temper leaders' willingness to use women in combat roles. By the early 2000s, GAM's lopsided fight against the Indonesian state had become a brutal full-scale civil war. During that period GAM faced rising resource constraints, and it increasingly looked to the international community for support in bringing a negotiated settlement to the conflict (Aspinall 2007, 253–254). Given its need for resources, GAM leaders arguably could have expanded the group's preexisting network of female members and deployed nontrivial numbers of women in combat roles. In addition to the thousands of women that actively supported the rebels, GAM had provided formal military training to hundreds more, who were deployed to gather intelligence, conduct reconnaissance, transport weapons to the front, and engage in acts of sabotage (Schulze 2003). Yet, even while the group's leadership used images of young women brandishing assault rifles as a form of "political theater" to draw international attention to the conflict, it generally excluded them from participating in combat (Aspinall 2009, 92–94; Barter 2014, 70; 2015, 349). This inconsistency is intriguing since it suggests that GAM leaders hoped to outwardly cultivate the belief that its forces included female combatants, but they were reluctant to actually send them into battle.

Pinpointing the specific motivations for the decision to circumscribe women's roles in the group is difficult. Yet, the traditionalist values of the organization and the community it relied on for support likely played an important role in this decision. GAM did not espouse an explicitly fundamentalist ideology. While the group traced its roots to the earlier, Islamist Darul Islam rebellion, the later incarnation fighting in the 1990s adopted a predominantly secular nationalist ideology (Aspinall 2007). Nonetheless, its leaders viewed Islam as a central component of Acehnese culture, and Islamic beliefs were highly influential in the development the group's political ideology. Moreover, GAM drew much of its support from rural parts of the Aceh, where traditionalist Islam was dominant (Aspinall 2009, 203–204). Given the nature of its support base, GAM's leaders likely perceived expanding women's roles as a potentially costly strategy that might alienate an important constituency that was deeply religious and held very traditional views on women's role in society. Consequently, despite the benefits GAM might have received from expanding women's participation in the group, the rebellion remained "almost exclusively a movement of men" (Aspinall 2009, 93).

Evidence of such costs is also visible in cases where rebels ultimately recruited substantial numbers of female combatants. For example, ZANLA's incorporation of female fighters and its nominal effort to expand women's rights in the areas under its control during the Zimbabwean War of Liberation generated resistance and resentment from male villagers and local leaders. While these adverse reactions did not fundamentally alter the group's reliance on female combatants, its leaders subsequently curtailed some of its domestic efforts to promote women's rights and equality (Kesby 1996; Kriger 1992, 194–196). Similarly, in Ethiopia, despite the large number of female fighters in the TPLF, its leaders eventually sought to constrain female recruitment, in part because male peasants and village leaders in some areas complained that the high rates at which women had joined the organization had destabilized domestic life in the countryside and upset the traditional social order (Young 1997, 180–181). These examples demonstrate that concerns over the social acceptability of female combatants can shape rebel leaders' decisions regarding female recruitment, even for groups that have already made the decision to mobilize women for war. In these cases, however, the leadership's commitment to social revolution (particularly for the Marxist-Leninist TPLF) and limited reliance on local patriarchal power structures to mobilize support likely muted the costs incurred by the recruitment of female combatants.[8] By contrast, groups that rely extensively on support from constituencies that hold deeply patriarchal attitudes are more likely to be sensitive to the potential costs of recruiting women and deploying them in combat.

Taken together, the previous discussions point to the combination of the intensity of the resource demands faced by a group and its leaders' (and base community's) underlying beliefs about the appropriateness of women directly participating in armed conflict as central factors in determining leaders' decisions about female recruitment. On the one hand, rebels can choose to ease their resource constraints by recruiting female combatants, but in doing so they risk incurring social or political costs for violating prevailing (and often deeply embedded) norms defining acceptable gender-based divisions of labor. Alternatively, they can forego female recruitment, which avoids concerns about violating gender norms but artificially constrains the number of potential recruits (i.e., supply) available to the armed group.

How rebels perceive the trade-off between expected resource gains and potential costs represents the central dilemma of female recruitment. Whether (and to what extent) the rebel leadership is willing to utilize female combatants therefore depends on both the severity of the resource constraints the group faces and the extent to which rebel leaders are sensitive to the potential costs associated with including them in the fighting force. This observation implies the existence of two types of rebel groups: those for whom recruiting women and deploying them in combat roles represents a nominal cost that is ultimately outweighed by the risk posed by rising resource constraints, and those for whom this action represents such a direct contradiction to the ideals of the movement or its community that the expected costs of the decision exceed any benefits the group might derive from expanding the pool of recruits to include female combatants.

War Costs, Ideology, and the Mobilization of Women for War

Above I described a generalized model of rebel recruitment decisions where female recruitment is a function of the group's overall demand for human resources, but where leaders' sensitivity to the costs associated with mobilizing women conditions the influence of those demands. In this section, I more fully develop this argument and specify the observable factors that determine a rebel group's demand for female combatants and its leaders' sensitivity to their recruitment. Specifically, I link the demand for female combatants to the overall severity of the conflict and the acute need for troops produced by periods of intense fighting, sustained troop losses, or rebel efforts to rapidly expand the rebellion. I also highlight the role of group political ideology in shaping leaders' willingness to transgress or challenge extant social norms and hierarchies and their baseline beliefs about the appropriateness of allowing women to participate in armed movements. I therefore conceive of ideology as a proxy for group leaders' sensitivity to the costs associated with utilizing female combatants. Ultimately, it is the interaction of these factors that determines the prevalence of female combatants in a rebel movement.

Conflict Severity and Resource Constraints

Successful rebel groups require substantial inputs of both human and material resources in order to achieve their broad political and military goals. As such, ensuring a continuing supply of recruits to send into combat—as well as troops to conduct the manifold other critical tasks necessary to maintaining the rebellion—becomes a central intermediate-term goal of any rebel movement. As the conflict intensifies, war costs mount, requiring increasing inputs of both human and material resources to sustain the movement. Failure to sustain the human resource needs of the rebellion increases the likelihood of rebel failure and defeat.

As with the case of the Yugoslav Partisans that introduced this chapter, evidence from a variety of rebellions suggests that the decision to recruit women and ultimately to train, arm, and deploy them in combat corresponds closely to periods of increasing conflict severity and acute human resource pressures. Leda Stott (1990, 27), for example, asserts that a shortage of male recruits in the early 1970s pressured the leaders of the Zimbabwean resistance movements to expand women's roles in the groups and to eventually deploy them in combat. Gisela Geisler's (2004) interviews with former ZANLA combatants and political leaders paint a similar picture of the "necessities of war" forcing the armed movement to reevaluate its position on training and arming female guerrillas and to its eventual decision to "broaden the revolutionary base" by promoting women's direct participation in the struggle. Similarly, rather than reflecting a feminist agenda or a commitment to equality, the dramatic increase in women's recruitment into the LTTE in the mid-1980s appears to have been a strategic response to the acute need for more fighters that resulted from the heavy losses sustained by the group in previous years (Alison 2003, 39; Herath 2012, 57, 77). The leadership of Sendero Luminoso appears to have come to a similar conclusion regarding women's potential contributions after experiencing mounting causalities during the escalation of its war against the Peruvian government (Cordero 1998, 352). Likewise, the intensification of the Vietnam War and the corresponding increase in U.S. troop levels after 1965 imposed severe resource demands on the Communist insurgency, which in turn encouraged its leaders to utilize women as fighters (Taylor 1999, 37).[9] Jocelyn Viterna (2013, 63) similarly describes the relationship between war costs and female recruitment in El Salvador in the following way: "The FMLN needed female guerrillas. Its high mortality levels early in the war created a personnel shortage that could not be filled by recruiting men alone." Finally, Margaret Poulos Anagnostopoulou (2001, 488–489) contends that acute manpower shortages in the Greek Democratic Army necessitated the mobilization (and eventual conscription) of women during the Greek Civil War.

Shelli Israelsen (2018) offers one of the most explicit analyses of the connection between resource demands and female recruitment. She argues that female recruitment closely maps to specific phases of civil conflict and contends that rebel groups are less likely to recruit women during the guerilla phase of the conflict, when the balance of power is highly asymmetrical and rebels routinely engage in hit-and-run tactics. However, these groups become comparatively more likely to recruit women during the later civil war phase of the conflict, when they are militarily more capable and better able to exert control over territory and populations. Her analysis of women's participation in the LTTE in Sri Lanka and the Karen National Union (KNU) in Myanmar provides support for this claim, demonstrating that in those cases the decision to deploy female combatants was a reflection of the intensity of the conflict and the need for troops. Importantly, her analysis also highlights the role that group gender ideology played in the decision, serving as a facilitating factor in the case of the LTTE and a constraining factor in the KNU. I return to the latter point below.

Resource demands also help explain one of the few cases of large-scale female mobilization to state military forces. As many as one million women may have participated in the Soviet Red Army and related irregular units during World War II, representing as much as 8 percent of the total force. Of these female soldiers, perhaps half a million reportedly served at the front, and 250,000 received military training (Campbell 1993; J. Goldstein 2001, 65–70; Pennington 2010).[10] According to Joshua Goldstein (2001, 70), extreme resource demands and the perceived existential threat represented by German advances played a central role in the state's decision to deploy women in frontline combat. Indeed, by the apex of the conflict the Soviet Union had mobilized some 90 percent of its male population in the war effort, suggesting that the country had literally run out of men to send to the front. While women had previously contributed to the war effort and many served in the military in noncombat roles, acute human resource pressures during 1942 and 1943 contributed to the Soviet leadership's decision to allow women to participate in combat roles on a voluntary basis.

Evidence from a variety of cases broadly supports the predictions of a demand-side model of female recruitment. All else being equal, as conflict intensity increases, rebel leaders become increasingly likely to create opportunities for women to participate in the movements as a means of ameliorating rapidly rising demands for human resource inputs. Yet, all is not equal across armed groups. As I elaborate in the subsequent section, ideology exerts a strong influence on groups' willingness to use female recruitment to address severe resource demands.

Group Political Ideology

Leaders' perceptions of the costs associated with female recruitment derive from their personal beliefs about the inherent incompatibility of women and armed combat, fears that introducing female combatants will diminish morale or sow discord in the ranks, and concerns that recruiting women will diminish support from the communities on which the group relies for other resources. The specific factors that generate and define these costs are often unobservable because they reflect the beliefs, values, and norms of the relevant actors or groups. The leaders of some rebel movements explicitly espouse a commitment to gender egalitarianism, which might signal that they either anticipate minimal costs from recruiting women or that by virtue of their convictions they are willing to tolerate such costs if they do occur. Jakana Thomas and Kanisha Bond's (2015) findings are suggestive of such a relationship, but they do not examine this question directly.

Yet, the absence of a formal position on gender equality is not necessarily indicative of a group's opposition to recruiting women and deploying them in combat. Moreover, some critics—both former female combatants and scholars—have asserted that the rhetoric of gender equality espoused by rebel groups rarely reflects a true commitment to egalitarianism and the advancement of women's rights (especially in the postconflict environment) (e.g., Kampwirth 2004; Nhongo-Simbanegavi 2000). Consequently, it is important to identify additional factors that reveal a movement's underlying tolerance for female combatants or the extent to which it perceives the deployment of women in battle as a potentially costly strategy.

In addition to a group's rhetorical commitment to gender equality, various group-level characteristics should reflect the baseline attitudes of the leadership and the group's constituents toward the inclusion of women in the group's combat force. Most importantly, the political ideology that a group espouses signals important information regarding its attitudes toward traditional social norms and hierarchies, including those related to gender roles and gender-based divisions of labor (Wood and Thomas 2017).[11] I follow recent studies and define *political ideology* as the package of ideas or beliefs that declares the grievances or aspirations of a particular social group, identifies a specific set of political or social objectives on behalf of that group, and proposes a course of action through which the group intends to accomplish those objectives (Gutiérrez Sanín and Wood 2014, 215).[12] More broadly, scholars often conceive of ideology as representing a shared set of inter-related beliefs regarding the appropriate order of society and how that order should be achieved (e.g., Jost, Federico, and Napier 2009, 309). Thus, ideology reflects not only the specific political objectives the group pursues and its strategy for attaining them but also its vision for the ideal type of social order it seeks to establish.

Because gender hierarchies and gender-based divisions of labor are prominent, and often fundamental, characteristics of systems of social order (e.g., Epstein 1989; Lorber 1994), the political ideology espoused by an organization reveals the attitudes, preferences, and beliefs regarding the appropriate roles for women within the model society endorsed by its leadership, members, and supporters. The specific political ideology a group adopts therefore reflects not only the rebel leadership's underlying personal attitudes and biases toward the deployment of women in combat roles but also informs its perceptions of the expected costs associated with this decision. Specifically, where women's participation in combat is perceived as a threat to the desired social order reflected in the group's ideology, leaders are likely to anticipate substantial costs from this decision and thus to resist employing women in combat roles as a strategy for managing resource constraints.

Women's participation in armed combat is often viewed as a threat to the extant (ubiquitously patriarchal) social order specifically because it represents a direct challenge to embedded gender norms and projects an image of female empowerment and liberation (MacDonald 1987, 3–4). For groups espousing orthodox, traditionalist, or fundamentalist ideologies, employing women in roles that directly contradict the traditional social order is both morally unpalatable to the leadership (and presumably to the rank and file as well) and risks alienating the constituencies whose interests the group claims to represent. Thus, the perceived costs associated with deploying women in combat are potentially quite high for the leaders of these groups.

By contrast, groups that seek to upend the traditional social order are likely to perceive comparatively few additional costs (and may perceive benefits) from the decision to employ female combatants—after all, they are already engaging in revolutionary action and have likely already decided that the backlash from conservative or reactionary elements of society are an acceptable cost to bear in pursuit of their objectives. Leaders of rebel groups whose ideology rejects or directly challenges traditional social hierarchies or explicitly promotes gender equality are therefore more likely to employ female combatants than leaders of groups whose ideology advocates preserving such hierarchies and supports traditional gender-based divisions of labor, which typically result in the subordination of women in society and their exclusion from traditionally male-dominated enterprises, including warfare.[13]

Because I am predominantly interested in ideologies that instantiate clear attitudes and beliefs regarding gender hierarchies and societal gender norms, I focus most directly on the opposing perspectives of staunchly secular (often antireligious), revolutionary leftist ideologies (e.g., Marxism, Leninism, Maoism) and orthodox or extremist religious ideologies (e.g., Islamism or Christian fundamentalism). Broadly speaking, these ideologies represent opposing poles on an ideological spectrum, particularly with respect to beliefs about gender norms and hierarchies.

Other ideologies, including variants of social liberalism, conservatism, fascism, occupy the space between the poles. For instance, groups embracing social liberal ideologies are more likely to support gender equality, exhibit more feminist attitudes, and challenge embedded gender norms than those embracing social conservatism, rightist, or orthodox religious ideologies. However, such groups do not typically advocate fundamentally remaking society or necessarily seek to subvert embedded social norms. By contrast, despite women's frequent participation in rightist political movements and armed groups, including groups adopting fascist ideologies, these groups have generally taken distinctly antifeminist positions with respect to women's roles in society and have often adopted positions that reflect essentialist beliefs regarding differences between men and women. For instance, fascist political movements in Europe in the early twentieth century frequently viewed women's primary role to be bearing and raising

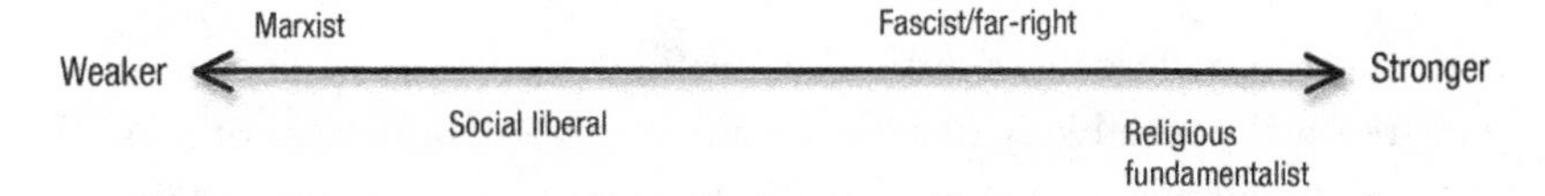

Figure 1.1 Commitment to traditional norms in selected political ideologies

children, advocated the subordination of women and their removal from the workforce, and sought to otherwise reinforce traditional gender hierarchies and roles (de Grazia 1992; Durham 1998). Similarly, right-wing extremist groups in the United States, including the Ku Klux Klan and Aryan Nation, count many women among their members but typically advocate traditional family structures and obedience to traditional gender norms (Blee 1996). Lastly, it is important not to conflate nationalism with right-wing extremist beliefs, despite the frequency with which they appear in tandem. Nationalism does not map easily onto the left-right spectrum and often intersects with and borrows from other ideologies. I therefore discuss it separately below. Figure 1.1 illustrates the placement of political ideologies on the spectrum of attitudes toward traditional gender norms.

A group's religiosity represents the primary dimension of its ideological orientation relevant to leadership's attitudes toward employing women in combat roles. I focus principally on this dimension because numerous studies have drawn strong connections between an individual's (or group's) level of religiosity and its attitudes and/or practices regarding gender equality and sex roles in society (e.g., Davis and Greenstein 2009, 94–95; Hertel and Hughes 1987; Inglehart and Norris, 2003; Peek, Lowe, and Williams 1991; Sherkat and Ellison 1999, 372). Overall, these studies conclude that individuals with greater levels of religiosity and those affiliated with ultra-orthodox or fundamentalist religious organizations are more likely to display in-egalitarian attitudes toward women than those who are secular or who are affiliated with less conservative religious groups.

Moreover, across different religions and cultures, fundamentalist beliefs and affiliations are positively related to support for patriarchal structures and adherence to traditional gender norms, both within families and in broader society (see Emerson and Hartman 2006, 135–136; Grasmick, Wilcox, and Bird 1990). Indeed, articulating and enforcing traditional gender roles represents a central concern of most groups that espouse fundamentalist ideologies (Beit-Hallahmi 2005, 28–29; Emerson and Hartman 2006, 135–136). Consequently, compared to secular political movements, those that embrace traditionalist, orthodox, or fundamentalist religious ideologies are much more likely to advocate for the (re)establishment of a social order in which men and women conform to traditional norms and where clear (male-dominant) gender hierarchies and divisions of labor exist.

RELIGIOUS IDEOLOGIES AND WOMEN'S PARTICIPATION IN POLITICAL VIOLENCE

The influence of religion on gender ideology should also be apparent in the ways in which women participate in armed resistance movements. Many women identify with fundamentalist and ultraorthodox social and political movements, and many are willing to mobilize on their behalf (e.g., Ben Shitrit 2016; Blaydes and Linzer 2008). The strength of their religious beliefs and support for the groups' social and political agendas help explain their willingness to engage in collective action in support of the groups. Moreover, at least some subset of women who hold fundamentalist beliefs are willing to support and actively participate in violent religious extremist movements. For example, Islamist groups such as Hamas, Hezbollah, and ISIS; the right-wing Hindu nationalist movement Rashtriya Swayamsevak Sangh; and violent ultra-Orthodox Jewish settler groups in Israel each count substantial numbers of women among their members. Despite women's prevalence in these groups, their roles typically conform to the traditional gender norms and hierarchies articulated in the movements' religious ideologies. This does not imply that female members of these organizations eschew violence—indeed, women have both supported and participated in violence on behalf of these groups in some instances. However, in the vast majority of cases, armed movements that embrace extremist or fundamentalist religious ideologies attempt to exclude women from roles that contradict the traditional patriarchal structures and gender norms embraced by the group.

Women's participation in militant Islamist movements illustrates the role religious ideology plays in shaping rebel leaderships' attitudes toward women in combat roles.[14] Like most other fundamentalist and orthodox groups, armed Islamist political movements generally reject the notion that a secular society can produce appropriate societal norms and instead believe that religious texts (the Qur'an and Hadith) should serve as the basis for social order. Consistent with these beliefs, they strive to reassert traditional social structures and patterns of social order, including the reinforcement or imposition of traditional gender roles and hierarchies (Beit-Hallahmi 2005; Moghadam 2003; Robinson, Crenshaw, and Jenkins 2006). The leaders of such movements are therefore likely to eschew the inclusion of female combatants because they perceive women's participation in organized violence alongside men as antithetical to their own beliefs regarding the appropriate social order. In addition, to the extent that the group's (male) fighters hold similar views, efforts to incorporate women are likely to be met with resistance and resentment, thus undermining group cohesion and impeding the movement's ability to achieve its objectives. A consequence of this commitment to traditional gender norms is that women are typically absent from military and political leadership roles and combat roles or appear in them only in small numbers and under exceptional circumstances.

The potential for backlash from a group's constituency represents an additional constraint on groups' willingness to utilize female fighters. For rebel groups that espouse orthodox religious ideologies, and are therefore likely to rely heavily on support from communities that hold deep-seated traditionalist attitudes toward women's roles in society, the use of female combatants represents a potentially costly strategy. As Nelly Lahoud (2014, 794) notes, the orthodox religious audience to which Islamist groups appeal for support is unlikely to support the large-scale deployment of women on the battlefield. Furthermore, Islamist groups have often undertaken substantial efforts to justify and legitimize the use of female suicide bombers to their conservative base communities (Gonzalez-Perez 2011; Davis 2013, 284; Von Knop 2007, 412).

That the leaders of these movements felt compelled to engage in these efforts suggests that the staunchly traditional communities on which Islamist groups rely for support have often been reluctant to accept this strategy, and it further implies that the large-scale deployment of women in combat would be met with resistance and possibly condemnation by its members. In other words, while this audience may tolerate—and even rally behind—the occasional female martyr, the acts of training, arming, and deploying substantial numbers of women in battle represent a direct challenge to the social order that this audience values and seeks to protect. The use of female suicide bombers, however, limits the overall number of women directly engaging in violence, thus allowing Islamist groups to frame their deployment as an exceptional strategy for exceptional circumstances rather than a broader, more consistent strategy that would potentially fundamentally alter the group's composition and its impact on society.

The disposable nature of suicide bombers also reduces the potential impact that violent, politically engaged women might have on traditional society. Because there are no female excombatants in the case of (successful) suicide bombers, there is no challenge in reintegrating them into society once the war ends and no potential pressures from groups of politically engaged women who might demand greater rights or opportunities as a result of their service to the cause (Davis 2013, 288). By limiting women's violent roles to suicide bombings, radical Islamist groups are therefore able to benefit from the strategic and tactical advantages they offer while minimizing the potential threats female fighters present to the groups' political ideologies and reducing the likelihood of significant backlash from their constituent communities. For these reasons, the overall prevalence of female fighters within Islamist rebellions is likely to be quite low.

While I have primarily discussed the influence of gender ideology on women's roles in radical Islamist groups, I anticipate that a similar relationship obtains in other religious fundamentalist rebellions. The general argument regarding the relationship between fundamentalist ideologies and the prevalence of female combatants should apply regardless of the specific religion of a movement. In other words, rebel groups motivated by fundamentalist ideologies, whether Christian, Muslim, Jewish, Hindu, and so on, are likely to enforce traditional gender-based divisions of labor within their ranks and therefore generally exclude women from combat roles.

MARXISM, WOMEN, AND VIOLENT REVOLUTION

Marxist-oriented ideologies and fundamentalist religious ideologies occupy opposite poles on the ideological spectrum, particularly with respect to the value they place on traditional social norms and hierarchies. While fundamentalist religious movements seek to protect traditional social hierarchies, rebel movements embracing Marxism (and its offshoots) typically seek to reshape existing social hierarchies. Moreover, the groups' gender attitudes are subsumed under this broader orientation toward social hierarchies. Fundamentalist movements openly pronounce their support for patriarchy and traditional gender roles. By contrast, Marxist-inspired groups often advocate (at least rhetorically) for greater gender egalitarianism and the expansion of women's rights.

Importantly, Marxist groups' rhetoric does not necessarily reflect a commitment to feminism as such. Rather, Marxist philosophy tends to view traditional social structures, including gender hierarchies, as an impediment to the liberation of the populace as a whole (Brown 2012, 12; L. Goldstein 1980, 331–332). Thus, for Marx, liberating women was not a central objective as much as it was part of a strategy of revolutionary liberation more broadly. Like Marx, Mao Zedong contended that equality for women represented a crucial step in the national revolution (Yuan 2005, 51–54) and identified women's subordination as one of the feudal-patriarchal systems that must be overturned. However, he believed that their liberation would naturally come as a part of the political and economic conflict in which they were engaged (K. A. Johnson 1983, 41). Nonetheless, Mao believed that the success of the revolution depended in part on the mobilization of women, and his recruitment strategy explicitly called for their inclusion in the movement (Gautam, Banskota, and Manchanda, 2001; K. A. Johnson 1983).

Subsequent rebel movements espousing variants of Marxist or Maoist ideologies have frequently adopted similar perspectives on the importance of gender egalitarianism and women's rights in furthering revolutions and similarly sought to more directly integrate women into the armed wings of their movements. However, consistent with their ideological roots, this effort had less do with feminism and more to do with the core principles of liberating all people from the tyranny of class structures and social hierarchies. Thus, in many cases the emancipation of women was seen as more a matter of economic independence, guaranteed by their integration into the workplaces of a new socialist society. Furthermore, for many Marxist-inspired movements, women represented a revolutionary force that could "neither be ignored nor allowed to pursue their own agenda" (Geisler 2004, 45–48). Many Marxist rebel groups therefore felt compelled to incorporate women because they were necessary for the success of the movement and because doing do was broadly consistent with the movement's ideology, not because of any deep commitment to feminism, which they often regarded as a bourgeois construction. For instance, while FARC's leftist ideology superseded the conservative patriarchal values of the society in which it existed, creating the opportunity to recruit women, its decision to do so was based at least partly on strategic considerations and the belief that recruiting female combatants was politically and militarily advantageous (Herrera and Porch 2008; Gutiérrez Sanín and Carranza Franco 2017). Similarly, Victoria Bernal (2000, 71) argues that the EPLF slogan "No Liberation Without Women's Participation," suggests that the group's leadership believed that victory necessitated women's support, and thus women's emancipation became a means to that end. The recruitment poster pictured in figure 1.2 reflects a similar sentiment.

Nonetheless, many Marxist and related leftist movements have made gender equality and women's rights core components of their political

Figure 1.2 EPLF poster (Ethiopia), published by the Eritrean Women's Association in Europe, 1980.
Source: Image courtesy of Basler Afrika Bibliographien

agenda. For example, in areas it controlled, Mao's revolutionary government passed laws that extended numerous rights to women, including suffrage, property ownership, equal pay, and freedom of marriage and divorce (K. A. Johnson 1983, 53–55).[15] While the EPLF leadership no doubt sought to instrumentalize women's liberation, it nonetheless made advancing women's status an important aspect of their fight against colonialism and the "backward, reactionary, and feudal elements of traditional society" (Bernal 2000, 66–67). Moreover, the group advocated gender equality and implemented numerous policies to advance women's rights in the areas it controlled (Bernal 2000; Coulter, Person, and Utas 2008, 10; Matsuoka and Sorenson 2001, 122). Similarly, the CPN-M directly challenged the traditional social, religious, and gender hierarchies of Nepalese society (Subedi 2013, 439–440). The group's leadership also made gender egalitarianism and the expansion of women's rights an explicit part of their platform and viewed women's participation in the revolution as a critical factor in its success (Gautam, Banskota, and Manchanda 2001; Manchanda 2004; Subedi 2013). The Marxist orientation of the leaderships of the FMLN in El Salvador and the Sandinistas in Nicaragua also led both groups to oppose patriarchal structures that limited women's opportunities and status in their societies and to propose policies aimed at promoting gender egalitarianism and advancing women's rights (Lobao 1990, 222–225; Kampwirth 2002; Reif 1986). As Norma Stoltz Chinchilla (1990) asserts, despite the frequent lack of attention to women-specific issues in orthodox Marxism, the specific political platforms adopted by these groups were largely consistent with modern Western feminism.

These examples illustrate the stated commitment of Marxist groups to challenging social hierarchies that restrict women's freedom and demonstrate the frequency with which they employ these perspectives in their rhetoric. This implies a close connection between the gender ideologies often present in Marxist-oriented rebel groups and the willingness of the leadership to utilize female fighters.[16] This contention dovetails with the empirical evidence that female fighters are more common in armed groups that espouse gender egalitarianism (Thomas and Bond 2015). However, two additional factors beyond Marxist leaders' potential commitments to gender egalitarianism that may impact their comparative acceptance of female combatants deserve discussion.

First, a common feature of all Marxist-inspired organizations is their goal of dismantling traditional social hierarchies and replacing them with a starkly different form of revolutionary order. Openly advocating such goals signals the group's willingness to absorb substantial costs for challenging the status quo. This cost tolerance should extend to any backlash anticipated from training and arming women to fight on behalf of the organization. The decision to deploy female combatants should represent only a marginal additional cost to those the group already incurs from its efforts to oust traditional village elders and minimize (if not outright demolish) the power of religious authorities, all of which are common objectives in Marxist rebellions. In other words, Marxist-inspired rebel organizations are likely to be less sensitive to issues associated with the recruitment of female combatants compared to armed groups espousing other ideologies.

Second, by virtue of their personal histories, leaders of Marxist rebellions are comparatively more likely than leaders of movements espousing other ideologies to possess positive assessments of women's capabilities, thus leading to them be more positively inclined toward women's participation in armed rebellion. Specifically, the leaders of leftist rebellions are heavily drawn from student political movements, trade unions, and other urban activist networks.[17] While women are often excluded from armed rebel movements, they routinely engage in high-risk collective action through their participation in these organizations. These contexts provide women opportunities to demonstrate their resolve, commitment, and skills to their male colleagues, which in turn help convince the leaders of Marxist rebellions that women possess the capabilities to participate in armed rebellion (Thomas and Wood 2018). As such, via their prior exposure to women who risked their lives for the cause in a nonviolent capacity, the leaders of Marxist rebel groups may be more positively inclined to accept women in the armed wing of the movement.

NATIONALISM AND OTHER IDEOLOGIES

The middle ground between these two poles of Marxist and orthodox religious ideologies is populated with a range of ideologies that place more or less emphasis on the importance of traditional hierarchies and norms in their ideal social order. For instance, groups adopting conservative or traditionalist ideologies short of religious fundamentalism are comparatively more likely to emphasize traditional social norms compared to those with liberal ideologies short of Marxism. Overall, I anticipate that the closer a group's ideology is to one of these poles, the greater effect the ideology exerts on the group's gender attitudes and thus its willingness to recruit women into the armed wing of the movement. However, I do not attempt to explicitly theorize on the specific effects of any particular ideology located within this intermediate space between Marxism (and related ideologies) and religious fundamentalism (and its various manifestations).

The one additional ideological type I briefly consider is nationalism. At its core, *nationalism* is the political principle that the political and national units should correspond to each other. Nationalist *sentiment*, however, is the compulsion to resist the violation of this principle and set it right again (Gellner 2008, 1–2). In terms of organizational goals, nationalist movements pursue independent states for their kinship groups and generally assert that adherents offer their loyalty to this group above all other affiliations (Van Evera 1994, 6). This includes both anticolonial self-determination movements and secessionist movements. Importantly, while nationalist groups advocate agendas based largely on group claims to autonomy or the superiority of an in-group over other groups, no broader sets of beliefs are common to all nationalist groups (Freeden 1998). That is, the defining element of nationalism is the pursuit of the nation, and nationalist groups frequently graft various components of other ideologies onto their core belief in national independence. Consequently, nationalism takes many forms and can reflect a range of additional ideological influences, including elements of Marxism, fascism, or religious fundamentalism.

Several scholars contend that nationalist agendas and ideals often serve to disempower, control, and repress women (Mayer 2002; Nagel 1998; Yuval-Davis 1993). However, because nationalist groups often import elements of other ideologies, the influence of nationalism on gender attitudes—and thus the group's willingness to contravene traditional social norms by recruiting female fighters—may depend more on the other ideological traditions on which the group draws than on its specific commitment to nationalism. For instance, many nationalist movements that employ substantial numbers of female combatants—including SWAPO in Namibia, Euskadi Ta Askatasuna (ETA) in Spain, and ZANLA in Zimbabwe/Rhodesia—incorporate aspects of Marxist or Maoist ideology into their political platforms, which likely influence their attitudes toward women's participation in the movement.[18] Other movements, however, infuse their nationalist agendas with religious ideologies. The Abu Sayyaf Group (ASG) and the Moro Islamic Liberation Front (MILF) in the Philippines blend fundamentalist Islamist beliefs with nationalist ideologies rooted in the goal of an independent Moro-Islamic state. Moreover, both limit women's participation in their respective organizations, and neither permits women to serve in combat roles.

As this discussion highlights, nationalist ideologies often incorporate traditionalist views of women's roles in society. However, in practice these ideologies often lack specificity in terms of their perspective on gender, and nationalist movements frequently incorporate elements from other ideologies into their own political platforms. These might include aspects of ideologies that adopt more expansive and progressive views of women's roles in society or those that are much more restrictive. Consequently, when isolated from other intersecting ideologies, nationalist ideologies are likely to have little direct influence on the prevalence of female fighters in an armed group.

Summary and Testable Hypotheses

This chapter presents a demand-side explanation for variations in the prevalence of female fighters across rebel movements. The argument focuses primarily on the factors shaping rebel demand for female fighters because rebel leadership, which is ubiquitously male-dominated, acts as the gatekeeper in the recruitment process and ultimately decides how it plans to recruit and whom. Thus, even where a substantial supply of women are willing to engage in high-risk collective action on behalf of a movement, rebel leaders determine how many women (if any) they will recruit into the group's armed wing as well as whether or not they will train, equip, and deploy them in combat.

I identified two factors that potentially help explain variations in the prevalence of female combatants across different armed resistance movements: growing resource demands and the group's political ideology. With respect to the former, I asserted that as rising war costs increase the demand for human resource inputs, rebel leaders are incentivized to increase recruitment in order to maintain the viability of the rebellion. While military leaders are initially reluctant to recruit women, the cost of failing to meet their human resource demands is an increase in the likelihood of the group's demise and a decline in the odds that it will achieve its objectives. I therefore expect female combatants to become more prevalent in a rebel movement as resource pressures become more acute. Given that most civil conflicts are characterized by asymmetric conflict, in which the government possesses a militarily superior combat force, rebels face persistent resource constraints. However, these constraints are likely to become more acute as the conflict intensifies. During periods of rapid conflict escalation, rebels experience greater numbers of troop losses and must more rapidly locate recruits to replace those troops killed or captured in the fighting in order to avoid defeat. I therefore hypothesize that the prevalence of female combatants are predicted by the intensity of the conflict. I formalize this claim in the following hypothesis:

H1: *The prevalence of female combatants increases as the severity of the conflict increases.*

The second factor I identified as relevant to determining the prevalence of female combatants in an armed group is that its baseline willingness to challenge extant social norms by training and deploying women to commit violence shapes the likelihood of female recruitment. Specifically, a group's ideology identifies its position on preserving or challenging existing social orders, hierarchies, and norms. This often extends to the group's commitment to prevailing gender norms and gender-based divisions of labor. As such, the group's ideology strongly influences the leadership's baseline tolerance for opposition to incorporating women into the organization's military wing. Similarly, ideology should determine the group's perception of the potential costs associated with utilizing women in combat. Overall, leaders of movements that seek to overturn rather than reaffirm traditional hierarchies—such as those groups espousing Marxist or Maoist ideologies—are more likely to tolerate the inclusion of female combatants and are less likely to fear repercussions from their constituents for doing so. While most Marxist-oriented groups are not explicitly feminist, many have adopted platforms and policies that at least rhetorically seek to provide women more equal status in their societies. I therefore hypothesize:

H2: *Female combatants are more prevalent among rebel groups espousing Marxist or other leftist political ideologies compared to groups adopting other political ideologies.*

By contrast, groups espousing orthodox and fundamentalist religious ideologies often seek to reestablish or reinforce traditional social orders in which clear gender hierarchies exist and in which traditional gender norms prevail. Leaders of groups that embrace orthodox and fundamentalist religious ideologies are likely to oppose the use of female combatants because it conflicts with their core beliefs about appropriate gender roles and the organization of society. In addition, they are likely to anticipate backlash from the base communities upon which they rely for support if they elect to train, arm, and deploy women in combat in substantial numbers. Thus, while the image of a woman carrying a Kalashnikov over her shoulder and a baby in her arms may be perfectly consistent with the social revolutionary ideology of Marxist movements, such images—and the values they represent—are largely anathema to orthodox religious movements. To be clear, many women play important and active roles in armed groups that embrace orthodox or fundamentalist ideologies; yet, the roles women play within these organizations largely conform to the traditional gender norms and hierarchies articulated in the movement's religious ideology. In other words, the leadership of armed orthodox religious movements attempts to ensure that women do not assume roles that contradict the social norms for which they advocate. I therefore expect that groups that espouse such religious ideologies will be highly unlikely to deploy women in combat roles.

H3: *Female combatants are less prevalent among rebel groups espousing religious political ideologies compared to rebel groups adopting other political ideologies.*

Finally, taken together, the arguments presented in this chapter suggest that the influence of ideology exerts a moderating influence on rebel leaders' willingness to accept women into the armed wing of the movement and deploy them as fighters as a strategy for managing acute resource demands. All else being equal, rebel movements become more likely to utilize female fighters as conflict-induced resource demands increase. Yet, as the discussion above highlights, not all groups respond to resource pressures in the same way. Rebel leaders that maintain a higher level of latent resistance to transgressing social norms or perceive greater costs from their base communities are less likely to respond to rising resource demands by recruiting women. In practical terms, rebel movements espousing orthodox religious ideologies are most resistant to the inclusion of female fighters and will typically resist doing so even in the face of severe resource constraints. This suggests an interactive effect between ideology and conflict costs.

H4: *The presence of a religious fundamentalist ideology conditions the influence of conflict severity on the prevalence of female combatants.*

CHAPTER 2

The Strategic Implications of Female Fighters

As I discuss in chapter 1, the leaders of most rebellions are reluctant to recruit women into the armed wing of the organization, and they are especially resistant to deploying them in combat roles. Even in the face of mounting resource demands, leaders' own gender biases as well as the fear of backlash from key constituencies dissuades many rebel groups from sending women into battle. Yet, some leaders eventually create opportunities for women to participate alongside men in the group's fighting forces, occasionally resulting in the large-scale recruitment, training, and deployment of female combatants on the front lines. For many of these groups, this decision represents a strategic response to acute resource constraints and reflects the leadership's ex ante beliefs about the effects of this recruitment strategy on the group's ability to mobilize essential human resources. Despite the centrality of this expectation in explaining the prevalence of female combatants, the strategic consequences of the decision to mobilize women for war are largely unknown.[1]

The dearth of research on this topic leaves open a number of important questions related to the impact of female combatants on the rebellions in which they participate. Most obviously, does extending recruitment opportunities to women succeed in alleviating the resource pressures that prompted the decision in the first place? Moreover, the recognition that female fighters typically represent a striking deviation from historical norms generates additional questions about the implications of their recruitment. For instance, how do local constituencies and external observers respond to the presence of female combatants? In what ways might the presence of female combatants shape the subsequent tactics and strategies the group employs? In order to probe these questions, in this chapter I therefore shift my focus from explaining the leadership's motivations for recruiting female combatants to understanding the strategic implications of this decision.

While the potential implications of women's recruitment are numerous, they have received minimal attention. I therefore initiate an exploration of this broad topic by investigating the manner in which female fighters influence rebel resource mobilization. I make two main arguments in this chapter. Consistent with the argument made in the previous chapter, I contend that the decision to utilize female fighters increases a group's overall mobilization capacity. In short, rebel groups that recruit female combatants maintain larger combat forces than those that do not. This occurs both because expanding recruitment opportunities to women increases the overall number of potentially viable recruits and because female fighters themselves often serve as important recruitment devices for rebel movements. Additionally, I argue that the presence of female combatants often assists rebel organizations in securing support from external actors. Drawing on previous studies of gender framing in the media and the strategic manipulation of gender narratives by political movements, I argue that the visible presence of female fighters draws international attention to the group's cause and positively influences external audiences' attitudes toward the group. These effects have downstream implications for the group's ability to secure resources from external nonstate actors such as diaspora communities and transnational activist networks. Ultimately, I argue that recruiting female fighters and highlighting women's participation in rebel propaganda and recruitment campaigns play important roles in rebel efforts to acquire the resources necessary to achieve their political and military goals.

Resource Mobilization, Support, and Rebel Success

Rebel groups seek to maximize their resource mobilization potential because the odds of achieving their military and political goals are heavily determined by their ability to do so. While a wide variety of observable and unobservable factors determine the ultimate outcome of a civil war, a number of previous studies link the balance of material capabilities between actors to a given group's probability of success (e.g., Clayton 2013; Cunningham, Gleditsch, and Salehyan 2009; Mason and Fett 1996).[2] Overall, a rebel group that minimizes the imbalance of capabilities between itself and the government, which generally reflects successful resource mobilization, is more likely to achieve at least a portion of its goals. The aggregate capabilities of an armed actor represent a combination of both military and nonmilitary resources. The former includes war materiel such as troops, small arms, artillery, and munitions while the latter includes financial resources, food and medical supplies, access to bases beyond the government's control, and the support and loyalty of the population. Because the survival and ultimate success of the movement is largely a function of its ability to mobilize these resources, efforts to acquire them represent a crucial proximate objective for rebel leaders.

The resources necessary to sustain the rebellion and whose acquisition shapes its outcomes originate from both domestic and international sources. As discussed in chapter 1, recruitment is a central occupation of rebel leaders. In order to survive and succeed, rebel movements require continuous inputs of material and human resources, which are most often obtained from the local population (e.g., Gates 2002; Leites and Wolfe 1970, 32–34; Weinstein 2007). Beyond serving as a vital source of recruits, local communities provide many essential nonmilitary resources to rebel groups, including food, clothing, camouflage, and intelligence. Indeed, loyal support from a base community is often essential to an armed resistance movement's ultimate success (e.g., Wickham-Crowley 1992; Wood 2003).

As a supplement (or a substitute) to domestically acquired resources, rebel movements often look to external actors to provide critical resources. Foreign governments represent the primary external source of military resources for rebel groups (Byman et al. 2001; Salehyan 2009; Sawyer, Cunningham, and Reed 2017). The provision of strategic assets by foreign states, which range from direct military intervention on the group's behalf to military arms and munitions, training, and access to bases on the sponsor's territory, often exerts a substantial influence on the trajectory and eventual outcome of a conflict. On average, groups that succeed in securing external support survive longer and are more likely to achieve their objectives (Balch-Lindsay, Enterline, and Joyce 2009; Gent 2008; Salehyan 2009). Yet, sovereign states are not the only sources of external support for armed resistance movements: transnational nonstate actors and international organizations routinely provide nonmilitary resources (and occasionally military resources) to armed resistance movements or their political wings. The effect of support from nonstate actors on conflict duration has received surprisingly limited attention. However, existing studies have often highlighted the benefits that rebel movements receive from transnational activist networks, diaspora communities, and international organizations (e.g., Adamson 2013; Bob 2005; Collier and Hoeffler 2004; Jo 2015).

I argue that, while it is largely overlooked in the existing literature, female combatants can play an important role in rebel efforts to mobilize critical resources. As such, recruiting female combatants can potentially deliver important strategic benefits to rebel movements. The argument presented in chapter 1 suggests that rebels recruit female fighters primarily because they believe doing so will help them fulfill domestic recruitment demands. The relaxation of acute resource constraints reflects the first—and most direct—strategic benefit the recruitment of female combatants provides to the rebel movements. Opening recruitment to women expands the pool of potential recruits, thereby increasing the total number of troops a rebel group can send into battle against the state. A second set of benefits derives from the manner in which the presence of female fighters in the ranks shapes observer attitudes and beliefs about the armed groups for which they fight. Specifically, the presence of female combatants—and particularly rebels' efforts to showcase and broadcast their support and sacrifice—directs international attention to conflict, legitimizes the group's goals, and garners support for the movement. These effects ultimately translate into strategic benefits by increasing the likelihood that the group can secure and maintain the support of nonstate external actors.

The first set of benefits is largely self-evident and directly flows from the initial strategic logic that led to women's recruitment. That is, recruiting women ameliorates resource demands. The effect of female combatants on the mobilization of additional resources requires a more nuanced explanation. To understand why female combatants exert a positive influence on external observers' willingness to support an armed movement, it is first necessary to discuss the importance of gendered beliefs about war and the use of gender narratives during wartime.

War and Gender Stereotypes Revisited

Hostility and violence are commonly viewed as predominantly male traits and associated with masculinity. By contrast, empathy, compassion, and vulnerability are more likely to be associated with women and femininity (see, e.g., Sjoberg 2010; Williams and Best 1990; Wood and Eagly 2002). Experimental evidence underscores the implicit association of men with threat, violence, and aggression (Rudman and Goodwin 2004).[3] Overt and implicit gender stereotyping of this kind serves to maintain and reinforce gender-based divisions of labor in society (Wood and Eagly 2002),[4] and the highly skewed gender distribution of direct participation in warfare is perhaps the most conspicuous and universal reflection of this pattern. In most societies, and throughout most of recorded history, combat has been viewed as an overtly masculine enterprise from which women have largely been excluded (J. Goldstein 2001). The historical record as well as implicit gender-based associations create the expectation that combatants are ubiquitously male, while women, to the extent they are present in conflict zones, are commonly assumed to be innocent victims of (male) violence (Carpenter 2005; Enloe 1999; J. Goldstein 2001, 304–305).[5]

An intriguing aspect of this discussion is that war can exert seemingly contrary affects on gender norms. On the one hand, war appears to reinforce traditional gender norms, pressuring men to assume combat roles and casting women as innocent victims. Yet, as Elisabeth Jean Wood (2008) observes, violent internal conflicts often fundamentally transform prewar social institutions, norms, and practices, including the roles of women in society. The social and economic disruptions caused by armed conflict can therefore create new opportunities for women to challenge traditional social norms, including through their direct and indirect participation in armed conflict (e.g., Kampwirth 2004; Mason 1992; Parkinson 2013). Moreover, as discussed in chapter 1, women's roles within armed groups tend to evolve and change over the course of the conflict. Contingent on group ideology and the strength of the beliefs held by relevant constituencies, women become more likely to participate in organized violence as the conflict endures, the intensity of the conflict escalates, and the resource constraints imposed on the rebels intensify.

Women's participation in organized violence may reflect a weakening of gender norms and muting of gender biases, particularly among rebel leaders.[6] However, it does not imply the wholesale rejection of such attitudes by the population of the conflict state or by external observers to the conflict. Gender norms that associate men with violence/aggression and women with peace/innocence are both widespread and deeply embedded. Thus, even where women take up arms on behalf of an armed group, these beliefs are likely to endure and continue to influence audience attitudes. The persistence of these beliefs also influences observers' expectations regarding the factors that motivate men and women to engage in violence. Men are assumed to embrace violence as a means of fulfilling their obligations to the state, achieving prestige, advancing ideological goals, or ensuring access to resources. Women, by contrast, are typically viewed as disinclined to resort to violence except under truly exceptional circumstances, such as protecting themselves or their children from mortal threats or unprovoked abuse. Indeed, the analogy of the lioness fighting to protect her cubs is commonly used to describe and justify women's engagement in violence (see Bayard de Volo 2001, 40–41).

This (assumed) distinction in motives for violence is critical to understanding both the ways in which women's participation in armed conflict is justified and how observers inside and outside of the conflict state perceive of this participation. Violence perpetrated by women is often construed as natural but uncommon and borne of sacrifice rather than duty or material pursuits (Cunningham 2009, 565). Appealing to the same underlying sociocultural structures that typically preclude women's participation in violence can therefore legitimize it during periods of large-scale conflict. Moreover, the perception that women's participation in organized violence represents self-sacrifice, and thus reflects truly dire circumstances, signals important information about the causes and the organizations for which they fight. Where women gain entry to the overtly masculinized world of organized political violence, their presence has two notable effects. First, it becomes noteworthy for its deviation from observer expectations and thus captures their attention. Second, it potentially influences observer beliefs about the nature of the conflict and women's motives for participation. As I discuss in more detail below, these gendered effects are apparent in media coverage of armed groups. Moreover, armed groups that utilize female combatants often exploit them for strategic gains.

Novelty and Framing in Media Coverage of Female Combatants

For many rebel movements, publicity is the key to survival and success. Often materially weak and operating in obscure or remote geographic locations, rebel groups derive substantial benefits from media coverage that draws international attention to the conflict (Bob 2005). Highlighting the presence of women in the movement represents one strategy through which rebel groups can enhance media coverage of their political struggle. Owing to the dynamics discussed above, the juxtaposition of images created by the presence of female combatants and their perceived contradiction to embedded assumptions about the nature of war promotes public interest and increases the salience of the conflict to external observers. Moreover, media coverage of female fighters differs in important ways from other reports of the conflict. Such coverage tends to humanize these women and rationalize their participation, thus framing them in a more sympathetic light. Both of these factors advantage rebel movements in their appeals to external actors and their efforts to portray themselves as legitimate political actors.

The perception of novelty surrounding female combatants is in apparent media coverage of both contemporary and historical conflicts. For instance, despite the general absence of women from GAM's fighting force, a 2002 article in the *New York Times Magazine* drew international attention to the group's "Widow's Battalion" (Marshall 2002) and to its separatist conflict in Aceh, Indonesia, more generally (Barter 2014, 70).[7] Similarly, during the Spanish Civil War national newspapers viewed the presence of female fighters in the Republican forces as a novelty, leading many outlets to publish photos of armed female fighters at rates that vastly outpaced their actual participation in the conflict (Lines 2012, 157–164).[8] More recently, the attention given to women's participation in the PKK and its allied Kurdish militias in Turkey, Syria, and Iraq (BBC 2014c; Robson 2014; Tavakolian 2015) and FARC forces in Colombia (Casey 2016; Miroff 2016; Tovar 2015) exceeds their actual prevalence within the ranks.[9]

The disproportionate attention journalists devote to female fighters may reflect their own attraction to the novelty of female fighters. Indeed, some scholars and non-Western journalists have recently criticized the Western press for its apparent fascination with "badass" women fighting in armed resistance movements in traditional societies (Dirik 2014; K. Williams 2015). However, as profit-motivated corporations, media outlets tend to feature stories that they believe will be of interest to their readers or viewers (see, e.g., Gentzkow and Shapiro 2010). The density of news coverage that female fighters receive (especially in Syria) therefore implies that Western viewers—not just journalists—express some interest in these stories.

Observer curiosity and interest in female combatants is also evident in the tens of millions of viewers who have watched any of the hundreds of videos available on YouTube and similar video-hosting websites that are devoted explicitly to women's participation in armed groups. For instance, a three-minute BBC (2014b) video report highlighting women's roles in the PKK has been viewed more than 1.7 million times. Documentaries produced by RT (formerly Russia Today) (2015) and VICE News (2012) focusing on Kurdish female fighters in Syria and northern Iraq have been viewed more than 2.5 million and 1.3 million times, respectively.[10] Dozens of other videos and news segments similarly documenting Kurdish women's participation in combat have each been viewed hundreds of thousands of times.[11] The conflict in Colombia has generally received less international media attention than the conflict between ISIS and the various Kurdish factions.[12] However, a *Guardian* (2015) video report largely focused on women's involvement in the FARC has received approximately a quarter of a million views on YouTube. Moreover, VICE recently produced a documentary entitled "Colombia: The Women of FARC," which highlights the long-running roles that female combatants have played in conflict and the potential roles they might adopt in the ongoing peace process (VICELAND 2016). Such coverage in both traditional and new media outlets suggests that female fighters—especially images of "girls with guns"—are capable of garnering substantial attention from outside observers.

In addition to generating audience interest, female fighters also shape the ways in which the media characterizes the groups on whose behalf they fight. Coverage of female terrorists and rebel fighters often employs gender frames and relies on entrenched gender stereotypes to explain the unexpected participation of women in armed political movements (see, e.g., Nacos 2005; Stack-O'Connor 2007a). As Patricia Melzer (2009) observes, gendered assumptions about terrorism typically preclude women's participation in it. While men are assumed to participate in political violence because of their ideological convictions or aspirations for power, the image of the female terrorist "warrants an explanation outside of political motives and/or personal power aspiration" (52). Thus, news coverage of women in political rebellions often intentionally highlights the exceptional and unexpected nature of their participation and looks to personal, emotional, or psychological factors to explain it. By focusing on female fighters, such coverage transmits important information about a group to relevant audiences.

Coverage of Kurdish female combatants in Syria illustrates how news reports intentionally or unintentionally cast the PKK and YPG in a more positive, sympathetic light. News stories and documentaries frequently characterize the female fighters in these groups as heroic and admirable individuals, risking their lives in a battle against a brutal adversary.[13] Moreover, these stories often use the presence of female volunteers in the Kurdish forces as a way to implicitly or explicitly frame the Kurdish forces and their cause as just and legitimate and use reports of ISIL's abuse and degradation of women to delegitimize that group. Mari Toivanen and Bahar Baser (2016) report that French media sources tended to highlight the emancipation and equality women achieved by fighting for the PKK and the YPG, while British sources tended to focus on the personal and emotional motives of the combatants, often framing them as victims who utilize the opportunities afforded by the Kurdish militias to redress their grievances and defend themselves from ISIL brutality.[14] In a similar manner, Laura Sjoberg (2018) contends that by highlighting the exceptional nature of female fighters and the extraordinary circumstances that drove Kurdish women to take up arms, news reports have tended to legitimize the Kurdish forces.[15]

Similar framing effects appear in media coverage of other civil conflicts as well. Moran Yarchi's (2014) analysis of the content of news articles published in American, British, and Indian newspapers about terrorist attacks in Palestine-Israel suggests that when the perpetrator was a woman, journalists were more likely to frame the events in ways that were more compatible with the Palestinian's preferred narrative about the conflict. Specifically, news coverage of these events focused comparatively more attention on the conditions in Palestinian society and on the group's motives for rebellion. Compared to stories about attacks committed by male terrorists, those involving female terrorists were more likely to provide detailed information about the perpetrator and to attempt to "contextualize and rationalize" her actions, thus humanizing her (681). Relatedly, a recent study of the depiction of (Arab and Jewish) female political criminals in the Israeli media found that journalists often employed stereotypical gender frames to explain women's unexpected participation in political violence. However, in so doing, these stories frequently highlighted events in the women's personal lives rather than their ideological convictions as motives for their participation in political violence (Lavie-Dinur, Karniel, and Azran 2015). Media reports of Chechen female suicide bombers have also tended to humanize the perpetrators and attempted to discern the motivations for their (unexpected) participation in the conflict (Card 2011). Journalists also appear to have adopted a more positive, sympathetic lens in their coverage of female LTTE combatants and suicide bombers (Stack-O'Connor 2007b, 48–49). The female guerrilla fighters of the Spanish Civil War likewise (initially) received positive coverage from the domestic and international press and were praised for their bravery, resolve, and generosity (Nash 1993, 275–276).[16] Consequently, news stories often appear to draw attention to female combatants and to humanize their actions and frame them as sympathetic actors. Such coverage can therefore benefit rebel groups attempting to establish the legitimacy of their cause to domestic and foreign audiences.

At first glance, framing female combatants as victims appears counterintuitive because they are, by their actions and the nature of their role as combatants, perpetrators of violence. However, this frame is effective specifically because of the power of embedded gender norms and assumptions that women do not normally participate in political violence. Because women's participation in combat is typically viewed as exceptional, their presence is often interpreted as a signal (or framed to suggest) that the conflict has become so dire that women are willing to transgress social norms or abandon traditional roles in order to defend themselves and their communities (Jacques and Taylor 2009; Sjoberg 2018; Viterna 2014, 199). Moreover, by providing women a forum in which to exercise political agency and defend themselves, the rebel groups for which they fight become the beneficiaries of this narrative of victimization, empowerment, and emancipation.

As this discussion suggests, there is evidence that media stories tend to frame women's participation in political violence in different ways than men's participation. Journalists often adopt highly gendered frames in covering or explaining women's participation. Yet, by emphasizing the exceptional nature of female combatants, contextualizing the motives for their participation, and highlighting the agentic and emancipatory opportunities afforded to them by participating in armed groups, media reports can produce a more positive, sympathetic depiction of the armed group than might otherwise emerge in the absence of female fighters. In the next section I consider how rebel groups may benefit strategically from similar effects and how some rebel groups utilize the presence of female combatants to advance a specific strategic narrative.

Constructing a Strategic Narrative

Many rebel movements are acutely aware of the potential impact that the presence of female combatants can exert on audience attention and perceptions of the movement. As with the framing devices employed by journalists and media outlets, armed resistance movements have often highlighted the presence of female fighters within their ranks in order to draw attention to the group and the grievances of the population it claims to represent and to garner sympathy and support (see, e.g., Nacos 2005). The decision to stress women's roles and participation often reflects one aspect of a broader effort by rebel leaders to construct a more positive, sympathetic narrative for their group and their cause. Particularly, rebel groups strive to cultivate a narrative in which they are sympathetic political actors with legitimate grievances and to counter the government's claims that they are simply a band of thugs, terrorists, or bandits.[17] This contest over legitimacy is central to rebel efforts to win support from both domestic and international audiences.

Striving for Legitimacy

Following recent literature in various social science disciplines, I understand legitimacy as "worthiness of support" (Lamb 2014, vi). When a population perceives that an actor and its goals or policies are legitimate, that population feels a moral obligation to support or emulate that actor. Establishing legitimacy is strategically (and practically) important to governments, political parties, activist networks, and social movements because it serves as the primary alternative to coercion as a means to ensure compliance (Horne et al., 2016; Zelditch 2001). In short, legitimacy induces support and participation, while illegitimacy fosters opposition and resistance (Lamb 2014).

These factors also help explain why rebel groups often strive to establish legitimacy in the eyes of key domestic and international constituencies. Compared to groups that suffer a legitimacy deficit, rebel groups that successfully convince these audiences of their legitimacy are in a superior position to secure resources and allies that can enhance their ability to achieve their primary political objectives. A key dilemma for rebels is that they often lack the historical or legal basis for legitimacy that governments and other established political actors enjoy. Rather, they must construct legitimacy from other sources, a process that often requires appealing to the beliefs and values of specific populations and undertaking efforts to "market" the movement to these audiences (see Bob 2005). Rebels adopt a variety of strategies and engage a range of actors in an attempt to foster a perception of legitimacy. These include appeals to ethnic traditions, participation in elections, securing representation in established political institutions, and gaining recognition by foreign states and international organizations (Jo 2015, 28–29; see also Huang 2016a). I expand the list of possible sources of rebel legitimacy by examining the manner in which gendered narratives and images can influence observers' beliefs about the legitimacy of a rebellion.

As the discussion in the previous section demonstrates, male and female combatants appear to provoke different responses from the audiences that observe and interact with them. Observers often view female combatants in a more sympathetic light and tend to believe that they are motivated more by rational fears or legitimate grievances than by ideological fervor, material gains, or inherent preferences for violence. While rebel leaders are often initially unaware of this tendency, which would preclude it as a motive for their initial recruitment, the leaders of organizations that have already opted to include women in the ranks are likely to become increasingly aware of this perception over time. For example, rebel leaders are likely to observe the responses of local civilians and international audiences to the presence of female combatants and update their beliefs about the potential strategic benefits associated with their use.[18] By explicitly highlighting women's participation in the organization, rebel groups hope to exploit these sentiments. Furthermore, to the extent that female combatants assist the group in establishing legitimacy, they may also assist the group's efforts to mobilize support from domestic and international constituencies.

In an effort to construct a more appealing and sympathetic narrative, some rebel leaders choose to intentionally highlight women's contributions and utilize images and descriptions of female combatants in their propaganda materials. Armed resistance movements in diverse regions of the globe have long recognized the powerful symbolism of women combatants and have intentionally incorporated these into their propaganda and outreach efforts (Loken 2018; O'Gorman 2011, 57; Stott 1990, 33; Weiss 1986, 147–148).[19] Particularly during the 1960s and 1970s, the image of the young female guerrilla fighter holding a machine gun—often carrying a baby on her back or in her arms—became an prominent image in the propaganda materials distributed by armed resistance movements and the solidarity movements and other allies that organized on their behalf (Bayard de Volo 2001, 42–43; Enloe 1983, 166; Jones and Stein 2008). For instance, the image of the female MK guerrilla became a popular mass image of the strong liberated woman in sub-Saharan Africa (Cock 1991, 167; Geisler 2004, 51). During the Vietnam War, the NLF used photos of "determined young women bearing arms and dressed for battle" in the propaganda materials it distributed internationally (Taylor 1999, 72). They also frequently used images of female guerrillas and descriptions of their heroic activities in their domestic propaganda materials and internal publications (76). SWAPO's propaganda publications likewise routinely included praise for its female combatants and photographs of female SWAPO fighters (Akawa 2014, 63).[20] These publications, most of which were printed in English, were intended to disseminate the ideals and objectives of the movement and to solicit international support (Akawa 2014, 66–67).

The image of the armed militia woman likewise became an important symbol for the anti-Fascist forces during the Spanish Civil War and was frequently used on propaganda posters printed by leftist parties and anarchist trade unions (Nash 1995, 50–51). Particularly, Lina Odena, a young female volunteer killed early in the conflict, became a Republican war legend. Her death—by suicide in order to avoid capture (and likely torture)—was widely reported in the left-wing Spanish press, and her image appeared on posters and in other propaganda materials (Lines 2009, 171–172). During the civil war in El Salvador, the FMLN and its sympathizers copied and distributed autobiographical poems written by Jacinta Escudos, a young

Salvadoran woman who had returned from Europe to fight for the FMLN. The El Salvador Solidarity Campaign in London later published these poems under the pseudonym "Rocio America" (Carter et al. 1989, 133). FARC has similarly utilized its female combatants in its public relations campaigns, including photo ops with international news media and direct interactions with the civilian population in the areas it controls (Herrera and Porch 2008, 614; see also Casey 2016; Miroff 2016; Tovar 2015). It also maintains a website that highlights the activities and contributions of its female members (BBC 2013).

Interestingly, these efforts often occur even where cultural norms strongly discourage women's participation in armed conflict. As the previously discussed case of GAM's political theater involving armed female supporters demonstrated, the international community rather than domestic constituencies represents the intended audience for many groups. For example, the Mujahadeen-e-Khalq (MEK) of Iran has emphasized the large numbers of female fighters within its ranks as part of a "charm offensive" aimed at legitimizing the group in the eyes of Western powers (Salopek 2003). While difficult to confirm, it seems this effort paid off, as the United States removed the MEK from its list of foreign terrorist groups in 2012. In addition, the websites of the various Kurdish factions active in Syria and Turkey, as well external organizations sympathetic to their causes, also explicitly highlight the prevalence of female fighters within the movements, draw attention to women's commitment to the cause, and emphasize the groups' support for women's rights (see PKK 2019; YJA-Star 2019; YPG 2017). As these examples suggest, many rebel movements intentionally highlight—and sometimes exaggerate—women's participation within the movement as an important aspect of their propaganda and marketing efforts.

The images featured here provide several examples of the ways in which rebel movements in various regions of the world have featured images of female combatants in their propaganda materials. Figures 2.1 through 2.4 were produced by or explicitly for a specific rebel movement.[21] Others, however, were produced by the groups' overseas activist networks or in cooperation with international activist organizations that supported the movements' goals. Figures 2.5 though 2.9 provide examples of images produced by such allied organizations. For example, figure 2.5 represents an example of the materials produced by the UK-based Namibia Support Committee (NSC), which raised awareness of the Namibian liberation struggle, lobbied Western governments on its behalf, solicited donations

Figure 2.1 PFLP poster (Israel/Palestine) by Marc Rudin, 1980.
Source: Image courtesy of the Palestine Poster Project Archives

Figure 2.2 NLF poster (Vietnam) by unknown artist, c. 1960s.
Source: Image courtesy of Dogma Collection Online

for SWAPO's political and humanitarian efforts, and transported supplies to the group's refugee camps. Like many rebel organizations, SWAPO had offices around the world and often worked with local advocacy networks and political organizations in those countries to disseminate information and solicit support. Figure 2.6 represents an image that was replicated widely, both in Rhodesia by ZANU and in Europe and the United States by pro-ZANU activist movements. Figures 2.7 and 2.8 represent similar

Figure 2.3 PIRA poster (Northern Ireland) published by Cumann na mBan, 1978.
Source: Image courtesy of Linen Hall Library Archive & Republican Publications, Belfast, Northern Ireland, UK

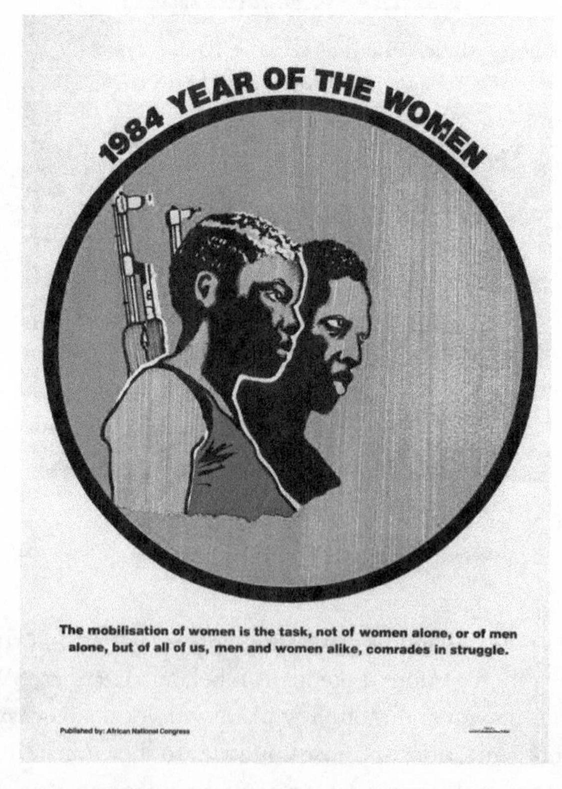

Figure 2.4 ANC poster (South Africa) published by the African National Congress, 1984.
Source: Image courtesy of African Activist Archive

Figure 2.5 Pro-SWAPO button (Namibia/Southwest Africa) produced by the Namibia Support Committee, United Kingdom, 1980s.
Source: Image courtesy of African Activist Archive

Figure 2.6 Pro-ZANU poster (Zimbabwe/Rhodesia) published by the Print Shop, NY, 1970.
Source: Image courtesy of Lincoln Cushing/Docs Populi

Figure 2.7 Pro-FRELIMO button (Mozambique) produced by the Liberation Support Movement, United States, 1972.
Source: Image courtesy of the African Activist Archive

Figure 2.8 Pro-MPLA poster (Angola) published by the Liberation Support Movement, Canada, 1972.
Source: Image courtesy of Lincoln Cushing/Docs Populi

Figure 2.9 Pro-FMLN poster (El Salvador) by Rafael Enriquez and published by OSPAAAL, 1984.
Source: Image courtesy of Lincoln Cushing/Docs Populi

depictions of female combatants in the materials produced by solidarity organizations working on behalf of FRELIMO and the MPLA, respectively. For instance, the North American–based Liberation Support Movement (LSM) helped coordinate support for and lobbied on behalf of a number of national liberation movements (predominantly in sub-Saharan

Figure 2.10 FMLN poster (El Salvador) published by Inkworks Press, CA, 1982.
Source: Image courtesy of Lincoln Cushing/Docs Populi

Africa) between 1968 and 1982. Figure 2.9 shows a poster produced by the Organization of Solidarity of the Peoples of Africa, Asia and Latin America (OSPAAAL) in support of FMLN guerrillas in El Salvador. This Cuban organization was founded in 1966 and was devoted to supporting (largely communist/socialist) liberation movements in the developing world. Finally, figure 2.10 shows a poster published in the early 1980s by the Berkeley-based publisher Inkworks Press for an FMLN office in San Francisco.[22] These images not only highlight the presence of female combatants in the armed political movements on whose behalf they were produced but also illustrate the ways in which networks of these organizations' overseas allies reproduced and transmitted these symbols to foreign audiences.

Gendered Imagery and Audience Attitudes

As the above discussion demonstrates, rebel movements have often incorporated images of women into their propaganda materials and highlighted the presence and participation of female fighters as part of their efforts to construct a more positive narrative. In order to accomplish this goal, rebels rely on the same entrenched gender norms and stereotypes as do the media (and other actors). They likewise apply similar combinations of gender frames in their efforts to strategically construct a narrative about women's motives for participating. Women's participation in warfare challenges assumptions about the nature of the conflict and leads observers to search for explanations rooted in the personal experiences that "drove" normally pacifistic women to take up arms. These explanations frequently focus on women's experiences of victimization and/or their desire to defend their children or their families from violence. In both cases, the implication is that women only take up arms in exceptional circumstances and in support of righteous causes.

The image of the female guerrilla—often depicted as simultaneously occupying the role of mother and combatant—conveys conviction, sacrifice, and a desire to protect the weak and the vulnerable (see Bayard de Volo 2001, 41–43; Macdonald 1987). This portrayal is an explicit attempt to legitimize their participation and, by extension, to legitimize the groups for which they fight. As Lorraine Bayard de Volo (2001, 41) asserts, the maternal image is directly tied to legitimate acts of protection: "the desire to protect one's children, even through the use of violence, was posed as a natural or divinely ordained maternal reaction [to threat]." MacDonald (1987, 14) similarly claims that the image of motherhood represents the "original protector/protected relationship." Such depictions, which are prominent in the images provided above, reflect a strategic effort on the part of the armed resistance movement to frame itself as a just defender rather than an aggressor (i.e., terrorist). By manipulating deeply embedded gender norms about warfare and conflict, the visible presence of female fighters helps the group to mobilize support and sympathy and legitimate the organization and its goals in the eyes of foreign and domestic audiences.

Rebels also utilize the victimization frame to promote a narrative in which they both defend innocent and vulnerable members of the population (including women) and provide them with opportunities to defend themselves against a brutal and repressive state. For example, the LTTE successfully used the presence of female fighters to highlight abuses committed by both the government and Indian peacekeeping forces, which they argued drove women into their arms in the first place, and to showcase the group's role in providing them protection and empowerment (Stack-O'Connor 2007b, 48). Jocelyn Viterna (2014) similarly argues that the FMLN intentionally drew attention to the high prevalence of female fighters within its ranks in order to construct a highly gendered narrative that portrayed the group as "righteous protectors" and defenders of the most vulnerable segments of society and to cast government forces as ruthless and unjust aggressors. Particularly, she asserts that the presence of female fighters helped the FMLN signal that the previous social order had been so thoroughly destroyed by state violence that women had no choice but to take up arms. During the Zimbabwean War of Liberation, women, including those who joined the movement, were portrayed as victims of the government's brutal counterinsurgency campaign, which guerrillas argued was a primary motive for their participation (Lyons 2004, 124). For example, ZIPRA was able to point to a 1978 government attack on a (predominantly) female training camp to vilify Rhodesian forces for attacking "innocent women" and gain international sympathy for their cause (125–126), despite the fact that these women were combatants rather than civilians.

Highlighting the motives for women's participation and employing the gender frames discussed above is intended to portray these fighters—and by extension the rebellions for which they fight—in a more sympathetic light. Ultimately, rebel groups and organizations allied with them utilize these images in an attempt to influence observer attitudes about the group and garner support for the movement and its goals from domestic and international audiences (see Bloom 2011; Nacos 2005; Sjoberg 2010). Viterna (2014, 199), for instance, asserts that the (perceived) innocence, vulnerability, warmth, and other traditional feminine characteristics displayed by young female FMLN fighters "tugged at the heartstrings of civilians" and increased their sympathy for the group's cause. Anne Speckhard (2008, 1000) similarly argues that the "Black Widows" generated public interest and even sympathy for the Chechen rebel cause.[23] FARC has likewise highlighted the presence of female fighters in the group during its public relations campaigns in an attempt to soften and humanize the image of the group (Herrera and Porch 2008, 614).

These frames can serve to justify both women's participation in the conflict as well as help legitimize the group and its goals. Previous studies suggest that some rebel groups recruit female fighters as an explicit strategy of bolstering the legitimacy of the movement (Bouta, Ferks, and Bannon 2005, xx, 13; Dahal 2015, 188; Shekhawat 2015, 9). However, as noted earlier in the chapter, many groups only realize the benefits of female fighters after making the decision to include them. Regardless of the motives underlying their recruitment, the presence of female fighters may serve as a signal to outside observers of the depth of the community's support for the movement and the group's resolve to achieve its goals (Alison 2009, 125; Dahal 2015). Demonstrating the breadth of community support and signaling that the movement is composed of "average" peasants pushed into these actions by extraordinary (and often extraordinarily unjust) circumstances represents one way that rebel groups attempt to challenge regimes' efforts to characterize them as being composed of radicals, extremists, and criminals.

While few studies have attempted to explicitly investigate the effects of female combatants on observer attitudes, some existing studies suggest that the gender composition of an organization can influence audience beliefs about its character and expected behaviors. Sabrina Karim (2017), for example, finds that increasing women's participation in the domestic security forces improved citizens' attitudes toward those forces. Specifically, citizens exposed to female officers expressed greater trust and confidence in the security forces and viewed them as more restrained and less likely to engage in abuse against citizens. One recent study similarly finds that among states emerging from protracted civil conflicts, national legislatures comprising greater numbers of female representatives are *perceived* to be less corrupt and more effective at delivering public services to their constituents (Shair-Rosenfield and Wood 2017). Analogous effects seem to exist in the corporate world as well. Specifically, gender diversity within firms' boards of directors increases observer perceptions of corporate responsibility (Bear, Rahman, and Post 2010; Rao and Tilt 2016). These studies, across a variety of contexts, suggest that gender diversity within an organization can positively influence observers' beliefs about its performance.

Gender also shapes observer perceptions of an individual's trustworthiness and propensity for criminal behavior. For instance, prospective voters tend to believe that female political candidates are less corrupt than their male counterparts (Barnes and Beaulieu 2014). Numerous studies have also found that judges and juries view female defendants and perpetrators as less blameworthy of their (alleged) crimes, less threatening, and less disposed to recidivism than their male counterparts, resulting in comparatively lower odds of incarceration and more lenient sentences for women (Daly and Bordt 1995; Rodriquez, Curry, and Lee 2006; Spohn and Beichner 2000). Importantly, this effect appears to hold in the case of politically motivated violent crimes as well (Bradley-Engen, Damphousse, and Smith 2009). Consistent with these findings, Carrie Hamilton (2007, 108–109) asserts that in Spain female ETA members were more likely to receive lighter sentences and endure less harsh treatment after arrest than were male members—a factor that incentivized the group to recruit female operatives. As these studies suggest, while women often participate in criminal activities and political violence (though at lower rates than men in both cases), they are often judged less harshly and viewed more sympathetically than their male counterparts. Given that audiences appear to be more sympathetic to female perpetrators, female combatants may serve to improve the overall attitudes of audiences toward the group.

The discussion above suggests that the presence of female combatants and the use of their images in rebel propaganda materials is intended to generate support and sympathy for the movement and to encourage mobilization. Rebels use female combatants to convey information about the group to both domestic and international observers. This is accomplished by highlighting the exceptional conditions that led to their participation (and corresponding transgressions of gender norms), framing women as victims-cum-empowered agents, and using them as a signal of the depth of the group's support in the community. Ultimately, these strategies are intended to help expand the movement and improve its ability to mobilize resources. In the next section I explain how these factors assist the group in achieving its political and military objectives.

Implications of Female Combatants

In chapter 1, I argued that attempts to address rising resource demands represent one of the principal motivations for rebel leaders to employ female combatants. By expanding the pool of potential recruits to include women, rebel leaders hope to increase the number of troops in their ranks, thereby increasing the odds that the movement survives and achieves its political goals. Above I discussed how female fighters sometimes become important symbols through which the movement attempts to shape the attitudes and actions of both domestic and foreign audiences. Specifically, the visible presence of female combatants serves to garner sympathy and support for the movement and its goals. This argument implies that rebel groups utilize female fighters in these ways because they *expect* that doing so will benefit them strategically. In this section I outline the specific channels through which these benefits accrue to rebels who utilize female fighters. I explicitly focus on the expected effect of female combatants on groups' ability to recruit and retain troops and their ability to secure support from external actors, principally transnational advocacy networks and diaspora communities.

Recruitment and Resource Mobilization

The most direct manner in which the decision to utilize female combatants benefits rebel movements is by increasing the number of troops an armed group can commit to combat. Provided that a substantial number of willing female recruits exist, the decision to allow (previously excluded) women to perform combat duties should expand the group's fighting force. Assuming that equal numbers of men and women are motivated to join the group and are capable of participating in combat, removing all restrictions on women's access to combat roles could double the size of a rebel movement's fighting forces. However, these assumptions are largely unrealistic, and both social and biological factors are likely to constrain the supply of female potential combatants.[24] Nonetheless, in some conflicts, large numbers of women take advantage of the opportunity to fight when it presents itself. Women eventually constituted between 20 and 40 percent of the combat forces of groups such as the EPLF, FARC, LTTE, and RUF, representing thousands of combatants. For these groups and others that counted a high proportion of female combatants in their ranks, failing to include women would have resulted in a substantial loss of viable combatants, thus potentially jeopardizing the movement's viability and diminishing the odds of it achieving its goals.[25]

In addition to their direct effect on the size of a group's combat force, female combatants may also indirectly benefit the mobilization capacity of the rebel movements in which they fight. They accomplish this in two ways, first through their ability to encourage or shame reluctant men to take up arms, and second by demonstrating to other women that participating in the rebellion is possible. Previous studies have thoroughly documented women's roles as successful recruiters for armed groups (e.g., Bloom 2011; Cragin and Daly 2009). Much of this literature focuses on the way in which women's participation as combatants increases pressure on men to participate by implicitly suggesting that men who fail to join the movement are less courageous or less physically capable than the women who risk their lives for the cause (De Pauw 1998; J. Goldstein 2001, 65; Stack 2009). According to Viterna (2013, 78), the FMLN viewed female guerrillas as the most effective recruiters because young men "felt shamed when young women in uniform and carrying guns entered their communities." The NLF also relied heavily on such recruitment strategies during the Vietnam War. As one young male recruit reported after encountering a group of female NLF guerrillas, "I am a man, and I could not be less than a woman. The pride of every man in our group was hurt. Consequently, all of us agreed to remain at the front" (Donnell 1967, 111). Gebru Tareke (2009, 91) recounts a very similar scenario in which female guerrillas of the Ethiopian TPLF encounter a group of young men and chide them into joining the rebellion by asking why the young men "are still at home with [their] mothers while the girls are fighting out in the fields."

Female combatants were reportedly one of the most effective methods of mobilizing ZANLA recruits during the Zimbabwean War of Liberation because they challenged men's feelings of masculinity and prodded them to participate (Weiss 1986, 80; Geisler 2004, 53). During the Guinea-Bissau Independence War, the Partido Africano da Independência da Guiné e Cabo Verde (PAIGC) similarly used female combatants as a recruitment tool: when female combatants visited a village, "all the men would join up so as not be shown up by the women" (Urdang 1984, 164). Likewise,

Serbian media and Serbian nationalist groups highlighted the patriotism and sacrifices of female combatants as a way to motivate reluctant men to fight in the Yugoslav wars of the 1990s (Kesic 1999, 188–189). The presence of female fighters can also push men to fight harder or endure greater hardships (J. Goldstein 2001, 201). For example, female FARC guerrillas apparently set a standard for ferocity and courage that male fighters felt compelled to meet (Herrera and Porch 2008, 614).

The presence of female combatants may also positively influence the recruitment of other women. Particularly in societies in which traditional gender norms prevail, a group's decision to deploy female combatants represents a tangible signal of women's empowerment and the ability to participate alongside men as nominal co-equals rather than being strictly relegated to less prestigious noncombat roles. This signal, in turn, increases women's willingness to actively seek out and take advantage of opportunities to participate in the rebellion when they are made available. Indeed, Cengiz Gunes (2012, 120) notes that for many young Kurdish women, encountering female PKK fighters "created a feeling of empowerment, which in turn created . . . a strong desire to join the PKK's ranks." Similarly, female guerrilla fighters played an important role in mobilizing the population in southern Africa rebellions because it was easier for female fighters to gain the support of other women (Geisler 2004, 53). During the Vietnam War, the NLF often used stories of heroic female guerrillas to inspire other Vietnamese women to participate in the conflict (Taylor 1999, 60–61, 78; Turner 1998, 37). The presence of female combatants (and leaders) in the FMLN early in the conflict in El Salvador challenged the stereotype that women were unfit for the rigors of guerrilla conflict and encouraged other women to join (Carter et al. 1989, 126), thus expanding the total number of recruits.

Finally, the visible presence of female fighters can encourage and strengthen support and loyalty toward the groups among local constituencies. The presence of female fighters helps accomplish this by enhancing the perceived legitimacy of the group's goals, garnering sympathy from observers and demonstrating the group's commitment to achieving its goals. Rebels that succeed in this endeavor are better able to secure the support of the domestic audiences that provide them with many of the resources essential to the rebellion's survival (Heger and Jung 2015; Mampilly 2011, 53–58; Viterna 2013, 2014). Thus, in addition to providing critical human resources, the presence of female fighters may bolster a resistance movement's ability to mobilize other forms of support from local constituencies, including the provision of food, shelter, and intelligence. Such support is often difficult to directly observe and quantify, but it should nonetheless serve to empower the rebellion and positively contribute to its resilience.

External Allies and Support Networks

As discussed at the beginning of this chapter, rebel groups often rely heavily on support from international organizations, foreign sponsors, and diaspora communities in addition to local constituencies. The ability to secure international support can have substantial implications for the group's survival and the ultimate outcome of the conflict. Rebel groups therefore often devote substantial efforts toward developing and maintaining external support networks that may include diaspora communities, transnational NGOs, and foreign governments (e.g., Coggins 2015; Huang 2016a; Jo 2015).[26] While strategic interests figure prominently in external actors' (particularly states') decisions about extending support to dissident movements abroad, movements that are better able to frame their goals as legitimate and their cause as just are more likely to successfully secure support (Coggins 2011). Propaganda serves a crucial role in rebel efforts to shape the beliefs and attitudes of external actors (Bob 2005; Jones and Mattiacci 2017).

As I have already discussed, images of female combatants figure prominently in the propaganda material of many rebel organizations, in large part because rebel groups that employ female combatants use such imagery as a way to signal important information about the movement. Principally, highlighting women's presence in the ranks serves as a strategy through which rebel groups seek to convince external actors of the legitimacy of their goals, counter the government's efforts to frame them as groups of violent extremists or thugs, and demonstrate the community's deep commitment to their cause. To the extent that these efforts succeed, directing attention to female fighters' presence in the movement and their contributions to the rebellion should enhance the group's ability to garner support from foreign and transnational actors. As I elaborate below, rebels target this gendered political messaging to a range of external groups and organizations. However, diaspora communities and transnational activist networks are most likely to respond to such efforts.

DIASPORA COMMUNITIES

Diaspora communities represent a critical source of material and nonmaterial support for many rebel and terrorist groups (Adamson 2013; Byman et al. 2001; Huang 2016b, 66–67; 2016b, 97). Paul Collier and Anke Hoeffler (2004), for instance, have asserted that diaspora communities' financial contributions to rebel groups at home represent one of the strongest predictors of civil war recurrence. The LTTE represents the most striking example of rebel group reliance on diaspora support, as the majority of its budget apparently came from remittances from the Tamil diaspora or from international investments (Byman et al. 2001). Other movements, including various Palestinian factions, the PKK in Turkey, PIRA in Ireland, and the Rwandan Patriotic Front (RPF) in Rwanda also depended on funding from diaspora communities (see, e.g., Byman et al. 2001, 41–42; Huang 2016b, 59). While the funding streams provided by diaspora communities are arguably meager compared to the value of resources provided by external states, they nonetheless amount to millions of dollars provided annually to many rebel movements and are often essential to the movements' survival.

Surprisingly little research systematically evaluates the factors explaining diaspora groups' support for insurgencies. However, much of the existing research points to the ability of the rebel movement at home to construct a salient connection with co-nationals abroad, establish the group as the legitimate representative of the population for which they fight, and engender sympathy for the movement and its goals (see, e.g., Adamson 2013; Matsuoka and Sorenson 2001). Many of the previously discussed mechanisms through which female fighters are expected to shape the attitudes and behaviors of domestic audiences should also influence overseas constituencies. For instance, members of diaspora communities often feel a profound sense of guilt for not participating in the armed struggles in the their homelands and thus look for other ways to assist (Matsuoka and Sorenson 2001, 95–96). Most often these efforts take the form of financial support, but they also frequently include political mobilization, networking with international NGOs and those located in the country of residence, and activism to raise public awareness among the population of the adopted country. Political entrepreneurs within the diaspora community working on behalf of the rebel movement are acutely aware of these feelings and often adopt strategic frames intended to enhance feelings of guilt, responsibility, or obligation in order to promote support among the members of the diaspora (Adamson 2013; Wayland 2004). Thus, just as female combatants "tug at the heartstrings" of the peasants in conflict zones and shame reluctant men into joining the rebellion, images of armed young women risking their lives for a cause are likely to spur support among constituents residing overseas.

Anecdotal evidence suggests that some rebel groups have successfully exploited the presence of female fighters in the ranks to mobilize support among the diaspora. For instance, the EPLF and its allied mass organizations staged events in diaspora communities in Europe and North America to celebrate the contributions and sacrifices of the *tegadelti* (EPLF guerrillas) fighting in Ethiopia, including its female combatants. Moreover, the EPLF-allied National Union for Eritrean Women (NUEW), which helped promote the group's women's rights agenda, was one of the primary mass-based organizations mobilizing resources among the diaspora community.[27] According to Atsuko Matsuoka and John Sorenson (2001, 122–123), many members of the Eritrean diaspora strived to emulate the sacrifices and self-reliance of the EPLF fighters. Most surprisingly, public awareness of the high proportion of female fighters in the EPLF reportedly exerted a positive influence on gender egalitarianism within the Eritrean diaspora, suggesting that this audience was not only aware of women's participation but were also influenced by it. By some estimates, the EPLF's highly organized network of support organizations was able to raise as much as $20 million per month from the global Eritrean diaspora (Clapham 1995, 88).

TRANSNATIONAL ACTIVIST NETWORKS

NGOs and solidarity movements represent another source of external support for rebel movements. These actors advocate on behalf of resistance movements—or their political wings—in foreign countries in order to raise public awareness for the group's cause, solicit donations, and pressure the governments of those states to intervene in the conflict diplomatically or to withhold support from the incumbent regime (Bob 2005; Clapham 1995; Sapire 2009; Sellström 2002). While these organizations rarely provide military support, they nonetheless represent an important resource for many rebel movements and have the potential to influence the trajectory of the conflict.

Numerous examples highlight the connection between external activist networks and rebel movements. During the Salvadoran Civil War, the Committee in Solidarity with the People of El Salvador (CISPES) carried out protests against U.S. funding for the Salvadoran government, called attention to mass human rights abuses committed by the regime, and sought to build support and solidarity for the FMLN's struggle.[28] In a similar fashion, organizations such the Chicago Committee for the Liberation of Angola, Mozambique and Guinea-Bissau and the Southern African Liberation Committee organized campaigns to raise awareness of and provide political and material support for liberation movements in sub-Saharan Africa. They also encouraged boycotts of companies conducting businesses in apartheid states and lobbied the U.S. government to withhold support from the governments of South Africa, Rhodesia, and other colonial states. In Mozambique, FRELIMO likewise received substantial support from numerous international charities and NGOs, including the Joseph Rowntree Fund of London, the Lutheran World Federation, and the World Council of Churches (Schneidman 1978). Left-wing student groups and parties in Europe and North America also routinely organized on behalf of armed resistance movements in the Global South—particularly those fighting colonial regimes—to gather humanitarian goods (e.g., food, clothes, blankets, etc.), raise funds, and increase public awareness for the movements' causes (e.g., Kaiser 2017; Kössler and Melber 2002, 109–111; Mazarire 2017; Sellström 2002). While the funding generated by these groups is often small in comparison to those provided to rebel movements by allied states, they nonetheless represent important inflows of much needed resources. Moreover, the lobbying and publicity efforts of such groups, which in some cases leads to diplomatic intervention by external states, is often at least as important as the tangible resources the solidarity networks can deliver.

While these groups were independent from the movements they supported, in many cases, the connections between transnational activist networks and the rebel movements they support are less clear-cut. The symbiotic relationships between anticolonial and anti-apartheid liberation movements and international solidarity efforts are well documented. NGOs in North America and Europe often desire to directly support liberation movements abroad and sometimes include political exiles from these organizations in their leadership (Sapire 2009; Saunders 2009). Numerous solidarity movements and NGOs have worked closely with rebel movements overseas, providing both direct assistance to the armed movement (or more commonly its political wing) and raising public awareness about the conflict. For instance, the NSC was an important source of overseas support for SWAPO and worked closely with members of the organization, even sharing a London office with the rebel organization until 1978 (Saunders 2009, 449). In addition, the Committee for Freedom in Mozambique, Angola and Guine (CFMAG), which was formed in the UK at the request of FRELIMO, worked with political parties, student groups, and churches to mobilize opposition to Portugal's colonial governments and to provide aid to the resistance movements challenging them (Bishopsgate Institute 2018). In the 1970s, various black solidarity movements organized to support anticolonial movements (particularly against settler regimes) in Africa and to attempt to shape U.S. policy toward the region. In 1973 alone the African Liberation Support Committee raised tens of thousands of dollars to support armed resistance movements in Africa, including ZANU, UNITA, PAICG, and FRELIMO (Cedric Johnson 2003, 490–491).

Connections with international governmental and nongovernmental organizations through their political wings also helped many armed resistance movements secure substantial amounts of humanitarian assistance from both NGOs and international organizations. For instance, numerous NGOs cooperated directly with the TPLF and EPLF's own relief agencies during Ethiopian Civil War. These agencies, which were connected to the rebel groups' governance structure, played a critical role in providing not only resources to the rebels but also publicizing the rebel groups' cause, legitimizing their struggle against the Derg regime, and assisting them in securing diplomatic contacts (Clapham 1995; Matsuoka and Sorenson 2001, 37). Similarly, SWAPO garnered substantial amounts of humanitarian aid from international solidarity organizations and multilateral agencies, much of which was intended for the refugee centers it managed in Namibia, Zimbabwe, and Angola (Akawa 2014, 77–78; Dobell 2000, 61–63).

The use of gender imagery in NGO campaigns is well documented (e.g., Carpenter 2005), and as the propaganda images above indicate, solidarity movements and NGOs allied with rebel movements frequently employ images of female combatants. For instance, SWAPO representatives and members of their overseas activist networks often highlighted women's roles and participation in the movement, including their roles as combatants, as a way to generate international attention and support (Akawa 2014, 77–78). Audiences located thousands of miles away and with little interest in the politics of the conflict state are more likely to pay attention to images that appear novel or unexpected and more likely to respond to evocative images such as those that depict young mothers taking up arms to advance their cause. As discussed above, such images rely heavily on the gendered expectations of observers that women's participation in combat is exceptional and must therefore signal the righteousness of the cause. Consequently, rebel groups that recruit female combatants and utilize their images in their propaganda materials are more likely to enjoy the support of substantial transnational solidarity efforts or foreign activism.

Summary and Testable Hypotheses

In this chapter, I have argued that female combatants strategically benefit the rebel groups that elect to recruit them by enhancing the group's recruitment and mobilization capacity, ultimately allowing it to field a larger combat force than it otherwise could have mustered in their absence. I identified two avenues through which this occurs. First, and most directly, the decision to open combat roles to women expands the group's potential supply of recruits. Even after accounting for the possibility that many women are not interested in fighting on the group's behalf and that some will not have the requisite physical capabilities to do so, allowing women to serve in combat should allow the group to recruit, train, equip, and field a larger number of combatants. Second, I argued that the visible presence of female fighters serves as an important recruitment instrument. Specifically, the presence of armed females in a group is likely to shame reluctant men into joining the armed struggle. I also argued that the presence of female combatants serves to expand the number of women willing to participate in the organization by demonstrating to other women that fighting with the rebels is a realistic option. Regardless of the specific mechanisms at work, the arguments broadly suggest that the presence of female combatants should increase the number of troops a rebel group has under its command. This provides the first testable hypothesis of this chapter:

H5: *Rebel groups with a higher prevalence of female fighters have larger combat forces than those with a lower prevalence of female combatants.*

A second set of arguments put forth in this chapter focuses on the potential role female combatants play in assisting the group's ability to mobilize international support. I highlighted the frequency with which rebel groups that include female combatants in their ranks utilize these women (or at least their images) in their propaganda materials and recruitment efforts. Rebels utilize female fighters and their symbolic value as part of their effort to construct a more positive image for the movement and counter the alternative, highly negative narrative (e.g., thug/terrorist) asserted by the government. In particular, the presence of female fighters in their ranks represents one strategy through which rebels attempt to signal the depth of community support for a movement, convey the legitimacy of its goals, and gain sympathy from external audiences. Establishing that observers in fact perceive female and male combatants differently and that the presence of female combatants enhances the perceived legitimacy of the groups that utilize them is central to demonstrating the validity of the arguments in this chapter. I therefore propose the following hypothesis:

H6: *The presence of female fighters increases observer support for the group and its goals.*

To the extent that female fighters successfully influence external audience perceptions of the movements that employ them, they may also increase the group's ability to garner resources from these audiences. While strategic interests typically dominate state decisions regarding support for foreign rebel groups and armed resistance movements, transnational actors are more likely to be swayed by impressions of a group's legitimacy or feelings of sympathy for its cause. By positively influencing external actors' beliefs about the legitimacy of the group and increasing their sympathy for its cause, female fighters improve the odds that the group receives financial or material support from diaspora communities, transnational advocacy networks, or other transnational constituencies. I therefore hypothesize:

H7: *Rebel groups that include female combatants are more likely to receive support from external nonstate actors compared to groups that exclude female combatants.*

In this chapter, I have presented a set of arguments regarding the potential strategic benefits associated with the decision to recruit female combatants. These arguments encompass both the direct benefits that come with relaxing constraints on the supply of recruits (which I argued in chapter 1 was a primary motivation for the decision to accept women into the ranks); the more indirect recruitment benefits that stem from the presence of female combatants, such as shaming reluctant men into participation; and the benefits associated with the manner in which the presence of female combatants assist the group in establishing legitimacy and garnering sympathy from relevant domestic and international constituencies. Ultimately, I argue, these dynamics benefit the group by allowing it to field a greater number of troops and helping it to secure the support of transnational actors. I empirically evaluate the veracity these arguments and the hypotheses drawn from them in the following chapters.

CHAPTER 3

Female Combatants in Three Civil Wars

How do the dynamics of female recruitment play out in actual civil wars? A potentially fruitful first step in evaluating the theoretical claims put forth in the preceding chapters is to trace the patterns of female recruitment in multiple contemporary rebellions. In the preceding chapters, I developed a series of arguments regarding the decisions of rebel leaders to recruit female combatants and deploy them in combat, and I explored the potential implications of this decision in terms of a group's ability to mobilize human resources and secure support from external actors. In order to probe the plausibility of these hypotheses, and to explore the dynamics driving female recruitment, I present a series of short case studies.

The three vignettes presented in this chapter focus on the PKK (and its allied militias) in Turkey and Syria,[1] the LTTE in Sri Lanka, and ZANLA and ZIPRA in Rhodesia/Zimbabwe. These cases are chosen in part to reflect sociocultural and geographic differences that might also account for variations in women's participation. The PKK's ongoing separatist rebellion, for example, is located in various contiguous Muslim-majority states in the Middle East, and the vast majority of its fighters are members of the Kurdish minority group. The LTTE's decades-long civil war took place in a small, Buddhist-majority state in South Asia, where members of the primarily Hindu Tamil minority engaged in a violent secessionist struggle against the Singhalese-dominated Sri Lankan government. Last, ZANLA's and ZIPRA's interrelated anticolonial rebellions occurred in a Christian-majority sub-Saharan African state under white minority rule. The memberships of both armed resistance groups were drawn from the majority black population of the country. However, ZANLA recruited more heavily from the Shona-speaking population and operated out of bases in Mozambique and Zambia while ZIPRA recruited primarily from the Ndebele-speaking population and operated almost exclusively from bases in Zambia.

Despite the geographic and cultural variations among the conflict settings, social norms related to women's roles and status in society are similar across the cases to the extent that, in each, women have historically been excluded from positions of power, have been viewed as subordinate to male family members (particularly fathers and husbands), and have experienced systematic marginalization in their access to social, educational, and economic opportunities.[2] Thus, in none of these cases does societal gender equality or women's level of political empowerment convincingly explain the leaders' decisions to recruit, train, and deploy thousands of women as guerrilla fighters.[3] Rather, prevailing social gender norms more likely served as strong deterrents against the large-scale mobilization of female fighters.

The vignettes included in this chapter are not intended to serve as direct assessments of the hypotheses presented in chapters 1 and 2. I rely on the quantitative analyses presented in the two following chapters for that purpose. These brief case studies are instead intended to illustrate the validity of the arguments put forth in the previous chapters and to highlight how various elements of the arguments apply to a set of a historical examples. While case-specific factors undoubtedly played important roles in determining the presence and prevalence of female combatants within the aforementioned rebel movements, the case studies highlight the importance of both resource demands and the group's ideological orientation in determining the leadership's attitude toward female recruitment. In addition, the cases help highlight the manner in which the leaders of the organization utilized the presence of female combatants to garner additional strategic benefits once women had been integrated into the movement. They therefore provide preliminary support to the primary arguments set out in the previous chapters.

The Kurdistan Workers' Party

The PKK, with an ideology deeply rooted in radical Marxist-Leninism, closely adheres to the expectations I outlined early in the book. In line with the arguments I put forth in chapter 1, the PKK boasts a high prevalence of female combatants: by most available estimates, they constitute more than a third of the PKKs combat forces (Marcus 2007). Women have been active participants in the PKK since the earliest days of the movement; however, the proportion of women in the organization was initially quite low, and female combatants only came to represent a substantial portion of the group's fighting force several years after the onset of the conflict. For instance, only two women, Sakine Cansiz and Kesire Yildirim (the wife of PKK leader Abdullah Öcalan), were reportedly present at the organization's founding meeting in 1978 (*France24* 2013; van Bruinessen 2001, 106).[4] By the time large-scale violence erupted in 1984, numerous female combatants served in the PKK, many of whom joined explicitly to escape the restrictions and expectations placed on them by traditional society (e.g., forced marriage) (Özeren et al. 2014; Tezcür 2015, 258). Yet the number of female guerilla fighters in the PKK appears to have only reached its current levels some time in the mid-1990s.

The combination of political ideology and resource demands largely accounts for the expansion of women's participation in the group over time. Consistent with the arguments presented in chapter 1, women's recruitment into the PKK was partly facilitated by the revolutionary ideology of the organization, which called for a radical (violent) reorganization of social and political institutions. Importantly, however, the following discussion also demonstrates the potentially endogenous nature of group ideology and its instrumental use in conflict. While the Marxist ideology of the PKK helps explain women's presence at the onset of the conflict and their inclusion (in small numbers) throughout the early years of the armed struggle, the leadership's (i.e., Abdullah Öcalan's) attention to gender egalitarianism and women's rights as an aspect of that broader ideology only becomes apparent later in the conflict. Specifically, it appears that the rapid expansion of the conflict in the early 1990s encouraged a sharp increase in the proportion of women combatants in the PKK.

Ideology and Egalitarianism

The leadership of the PKK has (at least rhetorically) made women's rights a central component of its political message, particularly in recent years. In one of the dozen or more essays available on the group's website that elaborates the PKK's perspective on women and gender equality, Abdullah Öcalan (2015a) declares that "the level of woman's freedom and equality determines the freedom and equality of all sections of society. . . . Liberated woman constitutes liberated society." In another essay, Öcalan (2015b) more fully articulates the underpinnings of this gender egalitarian ethos:

> I have often written about . . . the ability to divorce from the five thousand years old culture of male domination. The female and male gender identities that we know today are constructs that were formed much later than the biological female and male. Woman has been exploited for thousands of years according to this constructed identity; never acknowledged for her labour. Man has to overcome always seeing woman as wife, sister, or lover—stereotypes forged by tradition and modernity.

Both statements reflect a (rhetorical) commitment to women's rights and gender equality. However, what is most interesting about the latter is that it reflects more than a commitment to societal equality or a vague claim to support women's rights. Rather, it incorporates language and ideas that are central to contemporary notions of gender egalitarianism, particularly in its recognition of the distinction between the social construction of gender and biological sex. Moreover, it explicitly acknowledges the historical subjugation of women and insists that men must alter both their behaviors toward women and their core attitudes and beliefs regarding women's status in society. While such positions would not necessarily be seen as revolutionary in the context of most contemporary Western cosmopolitan societies, they are fairly striking given that they emanate from the leader of an armed movement located in a predominantly rural and underdeveloped region of the Middle East where traditional gender norms are deeply embedded.

While skeptics might charge that such statements are purely superficial and not indicative of any real commitment to egalitarianism, there is at least some evidence that the PKK has made a concerted effort to translate this rhetoric into policy and practice. Within the organization at least, it appears that rank-and-file members have increasingly internalized the leadership's gender egalitarian message. Both male and female PKK guerrillas appear to have become increasingly aware of the impact of traditional social structures on women's opportunities, become better able to recognize entrenched gender inequalities, and in some cases have begun to adopt the language of women's rights (van Bruinessen 2001, 105–106). In terms of policy applications, the PKK and its allied factions have imposed laws banning polygamy, child marriage, and unilateral divorce. Laws ensuring equal inheritance between male and female heirs have also been imposed in areas liberated by the PKK, and therein women now frequently serve in positions of political leadership (Bengio 2016, 38–39).

Paul White (2015, 146–149) further claims that since the early 1990s, the PKK has transformed itself into a "feminist organization" through encouraging women's leadership and the establishment of autonomous women's organizations at every level of the movement. It is also worth noting that PKK-allied Kurdish factions, such as PJAK in Iran and the YPG in Syria, have adopted similar positions on women's rights and implemented similar policies.[5] Consequently, the available evidence suggests that the group's Marxist ideology—albeit one that has been strongly shaped by Öcalan's own interpretations and beliefs—has exerted a powerful influence on the prevalence of female combatants in the movement as well as the positions that they fill and the status they enjoy. Yet, as I discuss in the next section, resource demands also help explain women's participation in the PKK. Moreover, the role of the group's ideology, particularly its attention to gender equality, in recruiting female combatants is not necessarily independent of the influence of resource constraints: the two factors jointly determined the growth in numbers of female fighters in the group in the early 1990s.

Resource Demands and Conflict Intensity

In chapter 1, I identified resource demands as a central factor explaining rebel leadership's decision to recruit and deploy female combatants, especially in large numbers. In the absence of a fundamentalist or orthodox religious ideology, rising resource demands incentivize rebel groups to recruit women and to deploy them in combat roles. The PKK's recruitment patterns largely conform to this expectation. While women had been present in the movement since its inception, there is little indication that the group included more than a handful of female operatives in its armed wing until the early 1980s. By that time, women routinely participated in the group's combat activities, though in fairly small numbers. However, beginning in the early 1990s, the gender composition of the PKK underwent a dramatic change, and the number of women fighting for the PKK rapidly increased (Tezcür 2015, 258). By as early as 1993, female fighters composed upward of a third of the rebel combat forces, and that number has remained remarkably consistent over time (Marcus 2007, 172–173; White 2015, 140–142).

While numerous factors potentially explain the rapid growth of female fighters in the movement's ranks, the timing of the change appears to correspond to a number of interrelated factors, including a shift in the PKK's recruitment strategy, which increasingly targeted university students and urban centers, as well as an increase in Öcalan's use of rhetoric related to women's rights and gender equality. Ideology, which is largely a static factor, would seem to be a poor predictor of the change in the rate of women's participation in the movement. However, it appears that Öcalan's commitment to gender egalitarianism—or at least his reliance on the rhetoric of gender equality—increased during the conflict. Notably, as the conflict lengthened and escalated, Öcalan increasingly made clear his belief that women's participation was vital to the success of the revolution (Celebi 2010, 108). Aliza Marcus (2007, 173), for instance, observes that the jump in the PKK's recruitment of female fighters coincides with Öcalan's more vocal stance on women's rights and his assertion that the liberation of women was a core responsibility of the revolution.

As with many rebel movements that include female combatants, women's issues were largely dismissed in the early years of the rebellion, and female members were generally treated as an indistinct component of the larger guerrilla organization. Despite the increasing presence of women in the PKK, their efforts to raise gender issues, discuss women's rights apart from national liberation, or organize a separate female movement were viewed by the leadership as potential threats to the group's cohesion. By late 1992, however, the leadership had constructed a separate women's organization within the PKK (Celebi 2010, 108).[6] This decision appears to have been made partly to address the interests and concerns of the group's growing number of female members. However, some observers maintain that expanding the group's recruitment capacity was Öcalan's primary motive for both the construction of a women's organization within the PKK and the impetus for more explicitly advocating the group's commitment to women's rights (Celebi 2010, 108–109). Thus, the decision may have been driven as much by strategy as by ideology. Though, given the presence of women in the organization in its earlier years, it would be difficult to dismiss these actions as strictly instrumental.

These changes in the group's rhetoric, recruitment strategy, and organizational structures regarding female members correspond to periods of increasing resource demands imposed on the PKK. While best characterized as a low-intensity conflict throughout most of the 1980s, by 1992 the conflict had escalated to a full-blown civil war, resulting in thousands of battle-related deaths per year. It was in the context of this rapid intensification of the conflict that the PKK's leadership actively embraced the large-scale recruitment of female fighters. This relationship is borne out by the data presented in figure 3.1. The figure presents the estimated number of battle-related deaths (y-axis) produced in each year of the conflict (x-axis) (Lacina and Gleditsch 2005; UCDP 2018). The dotted vertical lines that cross the x-axis at 1990 and 1993 represent the window of time during which most sources suggest that the PKK began to dramatically increase the number of female combatants within its ranks.[7]

Overall, the figure suggests that the timing of the increase in female combatants in the PKK corresponds to a window of time during which the costs of the conflict—and by extension the resource pressures faced by the PKK—were rapidly increasing. Specifically, the annual number of deaths attributed to the conflict increased from about eight hundred in 1989–1991 to more than four thousand in 1993, representing a five-fold increase. While not all of the causalities captured in the figure were borne by the PKK, the sharply increasing intensity of the conflict during the period likely imposed substantial resource demands on the PKK.[8] Most notably, given the high rate annual deaths, the group faced intense pressure to replace troops lost in combat in order to resist defeat at the hands of the Turkish military forces.

Figure 3.1 presents a general picture of the rising conflict costs during the window in which female recruitment by the PKK increased. However, a more detailed accounting of the conflict further highlights the correlation between rising resource demands and female recruitment. Beginning in the early 1990s, the number of confrontations between the PKK and the

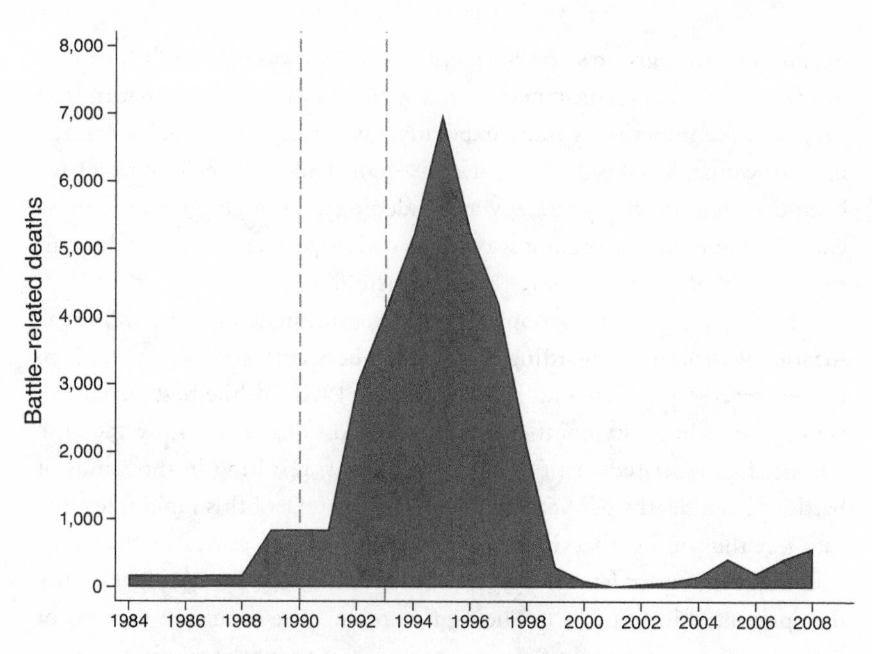

Figure 3.1 Annual battle deaths during the PKK insurgency
Sources: Data on battle deaths from Lacina and Gleditsch (2005) and UCDP (2018)

Turkish military rapidly increased, troop losses skyrocketed, and some of the group's units were forced to relocate outside of Turkey (Marcus 2007, 142). The group's ability to mobilize resources briefly increased in the early 1990s as a result of the combination of access to territory in northern Iraq following the Gulf War, a series of popular uprisings in towns along the Iraqi and Syrian borders, and growing Kurdish resentment of the brutality of Turkish counterinsurgency tactics. However, its power peaked by 1992, and thereafter the group lost ground to an expanding counterinsurgency campaign mounted by the superior forces of the Turkish military (Tezcür 2015, 259).

It was also during this period that PKK violence against civilians peaked. Based on data from the Global Terrorist Database (START 2016), there were roughly 80 terrorist incidents attributed to the PKK in 1989; however, that number rose to more than 350 by 1992.[9] Similarly, according to data available from the UCDP (2018), the PKK was directly responsible for approximately 70 civilian deaths in 1989, but that number rose to more than 200 in 1992 and to over 400 in 1993.[10] I reference these figures because the use of terrorism and intentional attacks on civilian targets is indicative of the resource stresses imposed on the PKK. Prior studies have repeatedly demonstrated that violence against civilian targets increases when rebels face increased competition for resources, declining military capabilities, or mounting troop losses (see Hultman 2007; R. Wood 2014). Coupled with the observation that the counterinsurgency campaign against the PKK was intensifying during this time period, it is reasonable to infer from the sharp increase in terror and anticivilian violence during this period that the PKK faced substantial resource constraints, and that these pressures provoked the leadership to consider new strategies by which to ensure its survival and expand the rebellion. That it was during this window of time that the PKK sought to dramatically expand women's participation in the movement suggests that it might have viewed this effort as one strategy through which to accomplish those goals.

As noted above, while present in the ranks from the outset, women constituted a fairly small number of the PKK's combatants in the 1980s. However, by 1993 they made up at least a third of the group's combat forces. The timing of the influx of female combatants into the PKK is therefore largely consistent with the predictions of the arguments presented in chapter 1. Specifically, resource pressures incentivized the group to search for alternative resource mobilization strategies in order to ensure group survival. The revolutionary Marxist ideology of the PKK and its stated objective of radically altering the existing social order, which partly explains the presence of women in small numbers in earlier stages of the rebellion, posed no deterrent to the leadership's willingness to meet their recruitment needs by incorporating larger numbers of women into the group's fighting force. Rather, the existing ideology facilitated this decision and likely created the opportunity for Öcalan to more explicitly highlight the centrality of women's rights and gender equality in the group's ethos. The explicit focus on gender equality, while perhaps present in some vague or nascent way early on, appears to have rapidly developed in the same window of time during which the group was recruiting larger numbers of female combatants. This implies that the ideological evolution may have been at least partly instrumental, reflecting Öcalan's efforts to mobilize more troops.

Media Attention, Propaganda, and External Support

The use of female combatants as a propaganda tool represented a primary component of chapter 2. As I argued, highlighting women's participation is often intended to attract greater attention from external audiences, increase audience perceptions of the legitimacy of the group's cause, and induce greater sympathy for the movement and its troops. Importantly, these factors are typically not drivers of the group's initial decision to recruit female combatants; rather, these potential benefits are generally realized only after the group has already begun incorporating women into its fighting force. Nonetheless, the decision to include women in the group's public outreach and propaganda materials represents a clear strategy intended to influence audience attitudes and assist the group's efforts to mobilize support and resources.

With the exception of a handful of other movements (e.g., the LTTE), the female combatants of the PKK have arguably received more international attention than almost any other female fighters. As I noted in chapter 2, the international media have devoted substantial attention to the "badass women" of the PKK and their participation in the group's conflict with ISIS. Based on the results of a search via LexisNexis, nearly four hundred English-language newspaper articles and newswire stories about the PKK (and its allied militias) published between 2010 and 2017 explicitly referenced female combatants in their titles—a rate of roughly fifty articles per year.[11] Notably, stories of Kurdish female guerrillas appeared not only in online and print news article but also in the pages of popular magazines, including fashion publications such as *Vogue*, *Elle*, and *Marie Claire*.[12] The prevalence of articles referencing Kurdish female fighters highlights the salience of the topic among (predominantly) Western readers.

The case of the "Angel of Kobane" stands out as a particularly illustrative example of the effect that images of Kurdish female fighters have exerted on external audience interest and attitudes toward the conflict. In late 2014 and early 2015, the Kurdish YPG and ISIS forces engaged in a series of brutal battles for control of the city of Kobane, which is located on the Turkish-Syrian border. During the months-long fighting, a series of stories and accompanying images of a young Kurdish female sniper who had reportedly killed more than one hundred ISIS terrorists emerged on Twitter and Facebook. The story of the heroic peasant-girl-turned-combatant named "Rehana" was shared thousands of times on social media and became a symbol of resistance to ISIS throughout the region (BBC 2014a; Del Re 2015, 91). Moreover, the image of the young female combatant and her heroic narrative focused the attention of tens of thousands of external social media viewers on the conflict in Syria.

The story's explicit focus on an individual female fighter and her personal story—including a discussion of her preconflict life as a law student in Aleppo and her father's death in the conflict—closely adheres to the gendered media narratives I discussed in chapter 1. It provides personal information about the subject and attempts to rationalize or explain the motives for her unexpected and seemingly exceptional participation in combat. As such, it is partly intended to (or may have had the effect of) garnering sympathy and attention for the fighter and, by extension, for the rebel group for whom she fights. Interestingly, as if to underscore the impact the narrative exerted on the conflict, ISIS's propaganda wing began (falsely) reporting that its forces had captured and beheaded Rehana, which in turn sparked a counter campaign by online supporters of the YPG to demonstrate that she was alive and well (Hall 2014; Varghese 2015). Coverage of this story in Western media further spread this narrative to a wider audience.[13]

The effectiveness of images of female guerrillas in attracting attention and soliciting support has not been lost on the PKK's leadership. The organization has long used the female fighters within its ranks as a way to encourage both domestic and international support. The presence of armed women among the contingents of guerrilla fighters that distribute propaganda material and recruit new members in Kurdish villages exerts a powerful influence on local observers. Consistent with anecdotes from other conflicts regarding the power of female combatants to boost male recruitment, reluctant Kurdish men were reportedly "shamed into action" when they realized young women were fighting with the PKK (Marcus 2007, 172). Young women, on the other hand, were often emboldened and empowered by the knowledge that women could directly participate in the rebellion alongside men. This projection of power had an enduring effect on some women, and many joined the PKK after meeting female fighters or seeing their images on PKK propaganda material (Gunes 2012, 120; Marcus 2007, 172–173; Tezcür 2015, 258).

The PKK and its allied militias in Syria and Iran have also routinely employed digital media, including their own websites, to explicitly highlight the prevalence of female fighters within the movement, showcase the sacrifices made by these fighters, and emphasize the groups' support for women's rights.[14] The official website of the PKK includes links to Öcalan's numerous essays on the topics of gender equality and women's rights, each available in English. The website of the YPG is also available in English and features numerous photos, videos, and stories of its female combatants. For instance, in one English-subtitled video posted on the website, a young woman named Delila recounts her experiences fighting for the group on the Rajo front, explains her motives for taking up arms, and asserts her (and her comrades') willingness to persevere in the fight for her people's freedom. Another English-subtitled video profiles a group of female fighters of the YPJ, an all-female PKK-aligned militia, battling ISIS in Raqqa, Syria. This video was originally produced by the international news agency Agence France-Presse and thus highlights the manner in which rebel groups often utilize sympathetic press coverage to help frame the movement and its goals. The website also includes a glossy eighteen-minute documentary-style English-narrated and subtitled video that explains the YPJ's association with the YPG and provides an overview of the group's goals and ideology. Most notably, it intentionally directs viewer attention to the group's focus on advancing women's rights and gender equality in the region and provides interviews with several YPJ members, who explain their motives for joining the movement and their understanding of the group's ideology and goals.[15]

These organizations also use social media platforms such as Facebook and Twitter to focus audience attention on women's participation. The Twitter account for the YPG (@DefenseUnits) currently has more than 60,000 followers, while the account for the YPJ (@DefenseUnitsYPJ) has more than 50,000 followers. On Facebook, the (seemingly) official page of the YPG Press Office counts more than half a million followers. As with the other information platforms used by these groups, the Facebook site devotes substantial attention to the presence and roles of the female combatants in the organization.

Sympathetic NGOs have also used traditional and digital media to highlight the prevalence of female combatants in the Kurdish forces. Among the most visible of the groups advocating on behalf of the Kurdish rebellions is an organization called The Kurdish Project. This nonprofit organization positions itself as a "cultural-education initiative" devoted to raising awareness about Kurdish culture among Western audiences.[16] However, a substantial amount of the information included on the group's website is devoted to the presence of female combatants in the various armed Kurdish organizations, particularly the YPG/YPJ. The organization also supports Females on the Frontlines, a website explicitly devoted to Kurdish female fighters.[17] The website is dominated by a short, highly polished video (with accompanying soundtrack) that is reminiscent of Western military recruitment commercials. The video closes with text asserting that the Kurdish female fighters bravely risk their lives each day to protect others, break gender norms, and fight for equality. The site also includes information on the prevalence of female fighters in the Kurdish forces (broadly defined), including a graphic comparing the proportion of the female combatants therein to the proportion of female troops in the national forces of several countries. Based on the statistics reported, the Kurdish forces contain the highest prevalence of female combatants in the world. A section of the website entitled "Meet the Women" includes photos, brief biographies, and quotes from various women purported to be members of the Kurdish forces. In addition, the website asserts that Kurdish female fighters routinely communicate with the world via social media and invites users to send a text or video message to these women via a hashtag provided on the page or through links available on the page.

The information presented on these websites is consistent with the arguments made in chapter 2 regarding the group's efforts to highlight the presence and participation of female members through its propaganda materials. As I argued, this effort is intended to assist the group in establishing its legitimacy and attracting the sympathy of external audiences. The PKK and its allied militias have made a concerted effort to attract attention and sympathy from external audiences. Particularly, the PKK's designation as a terrorist organization by the United States and European Union has placed substantial constraints on the movement. For this reason, it has for years endeavored to be removed from these governments' official lists of terrorist actors. The PKK's effort to highlight the presence of female fighters and its advocacy of gender equality and women's rights are, in part, intended to challenge the dominant narrative that it is a brutal terrorist organization and to promote a counternarrative in which it is a legitimate political actor that supports democracy, human rights, and women's rights.

Importantly, there is some evidence that this propaganda offensive is succeeding. First, it is worth noting that, despite its close connection to the PKK, the YPG has never been added to these lists of terrorist organizations. Second, numerous scholars, activists, and some policymakers in the United States and the European Union have called for their governments to reconsider the PKK's terrorist designation, particularly in light of its role in the fight against ISIS. A petition to remove the PKK from the U.S. terror list, posted on the White House website, gathered more than 33,000 signatures, and a similar position posted on the UK Parliament's website garnered 6,000 signatures.[18] While it is difficult to establish a causal link between the group's (and its allies') frequent use of women in its propaganda materials and the apparent support it has gained among international audiences in recent years, the general relationship is consistent with argument in chapter 2.

The Liberation Tigers of Tamil Eelam

While not as prevalent among the LTTE as among the PKK, female combatants nonetheless represented a substantial portion of the LTTE's fighting force. Estimates of the proportion of female combatants in the group vary, but most sources suggest that, at peak, female fighters constituted between 15 and 30 percent of the LTTE fighting force, with the former probably representing a more realistic estimate (Alison 2003, 39; Gonzalez-Perez 2008, 62; Stack-O'Connor 2007a, 45). As with the PKK, ideology partly explains the high prevalence of female combatants in the LTTE. Unlike the PKK, the LTTE did not overtly fuse Marxist ideology and nationalist beliefs. Nonetheless, it cultivated a revolutionary nationalist ideology that played an important role in facilitating female recruitment, particularly as the conflict intensified in the late 1980s. Moreover, as I discuss below, the LTTE was particularly adept at utilizing the presence of female combatants to garner support and sympathy from domestic and foreign audiences.

Nationalist Ideology and Women's Rights

Tamil autonomy from the Sri Lankan state represented the primary motive for the LTTE's rebellion, and this goal was reflected in the LTTE leadership's nationalist ideology. The group's leadership publicized the systematic political exclusion, social discrimination, and violence perpetrated by the Singhalese-dominated government against the country's minority Tamil population in order to mobilize support for national independence.

As is common among nationalist groups, the LTTE cultivated an independence narrative that drew inspiration from the (supposed) grandeur and power enjoyed by the Tamil community prior to its incorporation into the modern Sri Lankan state, and it incorporated symbols of Tamil history and culture in its political messaging and propaganda (Hellman-Rajanayagam 1994, 54–56).

While espousing a broadly nationalist ideology, the specific political belief structure of the LTTE was more complex. Velupillai Prabhakaran, the group's longtime leader, blended nationalist goals with a belief system that called for the revolutionary restructuring of traditional Tamil society. Notably, the LTTE's ideology called for the rejection and disestablishment of inegalitarian and oppressive traditional social structures, including the caste system and practices and laws that limited women's rights and opportunities (e.g., dowry and marriage laws) (Alison 2009, 126; DeVotta 2009, 1035; Hellmann-Rajanayagam 1994, 69–70). As with the PKK, some skeptics have claimed that the group's advocacy of women's rights was largely rhetorical and did not reflect a deep commitment to gender equality (Herath 2012, 57). Yet, in the areas the LTTE controlled, the group sought to implement policies that expanded legal protections and rights for women (Hellmann-Rajanayagam 2008, 7; Stokke 2006, 1027). In this sense, the LTTE's ideology of revolutionary national liberation shared some similarities with Marxist-inspired ideologies like those that guide the PKK. In spite of the occasionally leftist tone adopted by Prabhakaran, there is little evidence he was ever committed to Marxism (Hellmann-Rajanayagam 1994, 56, 68; Staniland 2014, 152).

The prevalence of women in secular nationalist movements like the LTTE underscores two points made in chapter 2 about the relationship between ideology and the gender inclusivity of rebel movements. First, a Marxist-inspired ideology is not a prerequisite for female combatants; however, the large-scale recruitment of female fighters likely requires a willingness on the part of the movement's leadership to challenge traditional social structures as well as the authorities that seek to defend them. This feature is largely absent in—and antithetical to—armed movements that espouse religious ideologies, which typically seek to preserve or re-establish such traditional structures. Second, and related, the role of women in the LTTE illustrates the ambiguous influence of nationalist ideologies on women's participation in armed groups. For groups that blend revolutionary beliefs with nationalist goals, the inclusion of female fighters is a natural extension of their efforts to reorder society. Yet, for nationalist groups whose ideologies focus on the preservation or reconstitution of traditional social orders, such as those that blend religious beliefs into their nationalist views, the inclusion of women would seem largely inconsistent, if not antithetical, to their beliefs. Consequently, nationalism as a broad class of ideologies is poorly predictive of women's participation.

While non-Marxist, the LTTE's revolutionary ideology partly explains the prevalence of female combatants in the movement. Indeed, both the LTTE's advocacy of greater rights for women and the inclusion of women in its ranks can be traced to the earliest days of the conflict (Hellmann-Rajanayagam 2008, 6). Like the PKK, however, the apparent intensity of the group's commitment to gender equality as well as the proportion of female combatants in its ranks increased over time. Thus, ideological factors most likely established the overall willingness of the movement to recruit women, but other factors account for the growth of women's participation over time.

War-time Resource Demands

In a similar fashion to the PKK and in line with the arguments outlined in chapter 1, resource demands appear to have played a significant role in shaping the LTTE leadership's decision to recruit larger numbers of female combatants. Conflict-induced resource pressures may also have deepened the group's rhetorical commitment to women's rights. Neil DeVotta (2009, 1035), for example, asserts that the gradual push for gender equality by the LTTE leadership corresponded to the group's need for fighters. More generally, several scholars have observed that the group's effort to recruit female fighters in substantial numbers began in response to the difficulties it faced in locating sufficient numbers of male recruits, which resulted in no small part from escalating violence and displacement in many Tamil-majority parts of the country (Alison 2004; Hellmann-Rajanayagam 1994, 70). This explanation for the increase in female combatants in the group conforms to the criticisms lodged by some feminist scholars who have sought to explain the LTTE's recruitment of women and position on gender equality as instrumental rather than ideological (Herath 2012, 57). In other words, these skeptics claim that the need for troops drove these decisions rather than any real commitment to gender egalitarianism and women's rights.

While numerous sources have linked the LTTE's increased reliance on female combatants to the group's need for additional recruits, it is useful to explore this hypothesis more thoroughly here. As noted above, women have participated in the LTTE since quite early in the movement; however, the extent of their participation has varied substantially over time but generally followed a trajectory of increasing integration. According to Margaret Gonzalez-Perez (2008, 61), the LTTE began recruiting women in 1979—before the onset of large-scale violence. However, prior to 1984 women were not included in the group's fighting force, and female recruits were largely confined to noncombat support roles such as nursing, logistics, and intelligence (Herath 2012, 171; Stack-O'Connor 2007a, 45). In that year, however, the LTTE created its first female combat unit, and it began training women at its bases in the Indian state of Tamil Nadu in 1985 (Alison 2003, 38–39).

While nontrivial numbers of female combatants were recruited into the LTTE as early as 1984, and women entered combat as early as 1985, the overall proportion of female fighters remained relatively low until the end of the decade (Alison 2003, 89; Alison 2004; Stack-O'Connor 2007a, 48). Given their fairly small numbers, these initial female combatants were often viewed as interlopers whose contributions to the group's goals were frequently questioned by male troops. As such, they felt constant pressure to prove their capabilities to the leadership and their male counterparts (Herath 2012, 171). There is, however, some indication that women's participation in the LTTE increased during the late 1980s. One such sign is the founding of the "Black Tigers" in July 1987 (Hopgood 2005). Female fighters have constituted a substantial portion of this elite, highly trained, and extremely lethal suicide force since its inception (see Herath 2012, 135–138; Stack-O'Connor 2007b, 52–53). During the same year that the leadership created the Black Tigers, it also established the first all-female training camps in Jaffna, and within two years the women's unit had its own leadership structure (Alison 2003, 39; Stack-O'Connor 2007b, 50). This suggests that the LTTE's leadership increasingly recognized the strategic value of female combatants during this period and had begun to see them as an integral part of the organization's fighting force.[19]

Beginning in 1990 or 1991 the proportion of female fighters in the LTTE dramatically increased (Alison 2003, 39; Gonzalez-Perez 2008, 61; Schalk 1994, 166–167). In addition, the number of women promoted to more senior roles in the organization and the rate of women's participation in the Black

Tigers increased after 1991 (Hopgood 2005, 67–68). By the early 1990s, women had become so common in the LTTE that they became viewed as a "normal part of the revolutionary family" (Herath 2012, 171). It is from this period (and continuing until the insurgency's demise) that women constituted upward of a fifth of the group's combatants.

The advent of female fighters in 1985, the expansion of their roles and numbers in the late-1980s, and the substantial increase in their numbers around 1990 correspond to periods of sharp increases in the intensity of the conflict and increased resource demands imposed on the group at those moments in time. The variation in the intensity of the conflict is demonstrated in figure 3.2, which presents the annual number of battle-related deaths during the Sri Lankan Civil War between 1984 and 2008.[20] As with the PKK data in figure 3.1, the dotted vertical lines represent the window of time during which women began entering the LTTE in large numbers. The timing of the initial wave of LTTE female combatants corresponds to the first major escalation of violence in 1984. As the conflict moved from a localized insurgency in the late-1970s and early 1980s to a full-scale civil

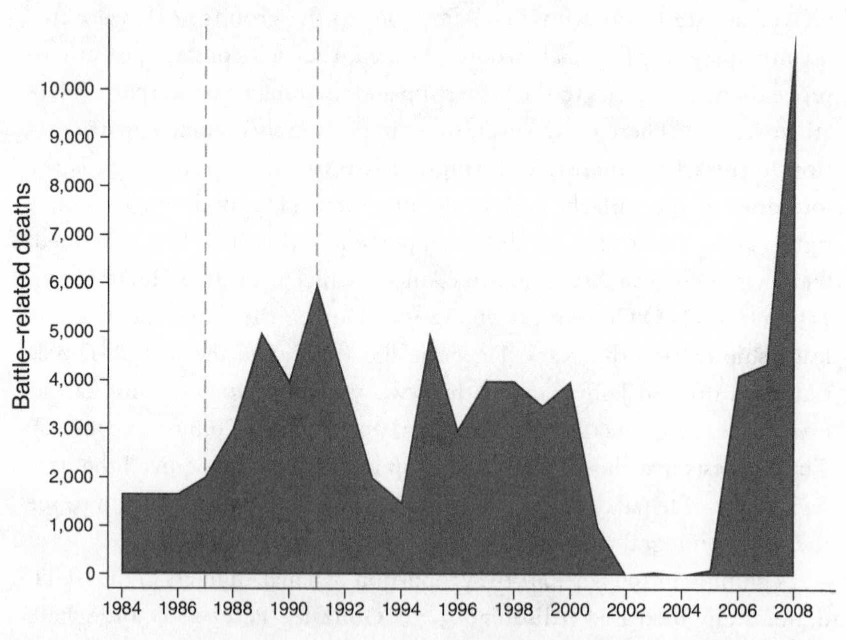

Figure 3.2 Annual battle deaths during the Sri Lankan Civil War
Sources: Data on battle deaths from Lacina and Gleditsch (2005) and UCDP (2018)

war in 1984, small numbers of female LTTE members began to occupy combat roles alongside their male colleagues. Based on the information presented in figure 3.2, the first substantial escalation in the conflict since the outbreak of large-scale violence occurred around 1987. That the group began to recruit and train more female combatants in 1987–1988 and chose to create the Black Tigers in that period is consistent with the resource demand hypothesis forwarded in chapter 1: as the conflict escalated and resource demands increased, the leadership encountered strong pressures to adopt new recruitment strategies in order to sustain the movement.

Not only was the conflict entering a more intense phase during the late 1980s, which required the LTTE to mobilize more troops in order to avoid defeat, the group also suffered numerous setbacks that required it to develop new war strategies and to redouble its recruitment efforts. The LTTE enjoyed substantial support from India beginning in mid-1983; yet, that support evaporated in 1987, and the Indian government instead redirected its support to the rival Eelam People's Revolutionary Liberation Front (EPRLF) (Staniland 2014, 155, 165–166). This change weakened the LTTE by depriving it of critical resources; worse still, it strengthened one of the group's primary competitors for recruits and resources. Relations between the Indian government and the LTTE further eroded following the arrival of the Indian Peacekeeping Force (IPKF) in 1987. As the IPKF attempted to disarm the militant Tamil forces, the LTTE launched a series of attacks against the peacekeepers. This opened up a new front in the conflict that proved extremely costly for the LTTE. In 1987 alone, the LTTE lost almost 10 percent of its forces (Stack-O'Connor 2007a, 47). By the end of the year, the IPKF had virtually expelled the LTTE from Jaffna; by late 1989, Indian troops had pushed the Tigers out of their remaining urban strongholds and into the jungle (Staniland 2014, 165–166).

At the same time that the LTTE was losing ground to the IPKF, the capabilities of the Sri Lankan military were rapidly expanding as a result of the implementation of a draft a few years earlier. Moreover, the LTTE, and Tamil militant groups more broadly, experienced persistent recruitment challenges resulting from the government's harsh counterinsurgency tactics, which targeted Tamil males for detention and interrogation and generally depressed the supply of available male recruits (Stack-O'Connor 2007a, 48). The growth of female fighters in the LTTE in the late-1980s therefore appears to correspond to a period of intensifying human resource demands and likely represented a strategy through which the group attempted to meet its recruitment demands.[21]

The IPKF withdrew from Sri Lanka in 1990, but the three-year, multisided war had depleted the LTTE. The withdrawal of Indian troops created opportunities for the LTTE to expand (or re-establish) their control over Tamil areas in the north of the country. However, during this period they also faced competition from the EPRLF, which had been trained and armed by the IPKF during its years in Sri Lanka. The EPRLF exercised control over portions of the east of the country and had rapidly expanded the number of troops under its command through a mass conscription campaign in the waning days of the IPKF's presence (Staniland 2014, 164–168). While the LTTE managed to dismantle the EPRLF's military operations by early 1990, the intracommunal fighting was brutal and bloody, requiring the LTTE to mobilize additional resources during a period in which it had already face substantial resource demands just to ensure its survival.

The LTTE emerged from this period battle-hardened, led by a skilled and experienced leadership, and largely uncontested as the militant force of Tamil nationalism. However, it was also depleted and eager to acquire the resources to expand its military operations. A ceasefire between the government and the LTTE collapsed in 1990, and by July 1990 a second phase of the war, commonly known as Eelam War II, was underway. The conflict escalated rapidly, resulting in nearly six thousand deaths in 1991—a 50 percent increase from the previous year. During this period the LTTE rapidly expanded its military apparatus and its state-building efforts. The rapid expansion of female combatants represented an important part of that effort.

Media Attention and Diaspora Support

As with the PKK, the female combatants of the LTTE received substantial attention from the media. Also like the PKK, it appears that the LTTE's leadership was acutely aware of the effects armed women could exert on audience attitudes toward the movement and its goals. As such, it sought to highlight the presence of female combatants in the organization and their sacrifices as a way to engender support and sympathy and to establish the legitimacy of the movement. The LTTE's efforts to highlight women's contributions to the group were also intended to promote support from the Tamil diaspora community, whose financial contributions were critical to the LTTE's twenty-five-year insurgency. As Alisa Stack-O'Connor (2007a, 48) argues, highlighting the inclusion of women allowed the Tigers to signal to both their domestic and international audiences that the group was representative of the entire Tamil nation and therefore more legitimate than its more exclusive and ideologically narrow competitors. Similarly, Miranda Alison (2009, 125) contends that the LTTE employed female fighters in part to demonstrate that the organization was an all-encompassing mass social movement rather than a violent extremist group. The leadership therefore sought to use the image of female combatants—and their unexpected and presumably exceptional participation—to soften the group's image and construct a more legitimate narrative than the one promoted by the Sri Lankan government.

Where most rebel movements appear to recognize the strategic value of female combatants after the fact, the leadership of the LTTE was acutely aware of the propaganda benefits of female combatants from quite early in the conflict (Brun 2005, 66–67; Stack-O'Connor 2007b, 97). For instance, the group's leadership produced a website that included accounts of the history and activities of its female guerrillas (Cragin and Daly 2009, 46). In addition, beginning in the early 1990s, the LTTE published the magazine *Cutantirap Paravai*, which highlighted women's role in the liberation struggle (Hellmann-Rajanayagam 1994, 69–70). The leadership also invited the press to interview select female combatants with the intention of allowing the media to juxtapose the personal stories of the female fighters with the abuses committed by the Sri Lankan government and the IPKF (Stack-O'Connor 2007a, 48). In line with the observations in chapter 2, female LTTE fighters appear to have frequently received positive and sympathetic coverage in media accounts. As noted in other conflicts involving female fighters, media coverage of women in the LTTE was apparently rarely negative and tended to portray female combatants and suicide bombers as victims rather than violent killers (Stack-O'Connor 2007a, 48–49). Even though the LTTE had become notorious (both domestically and internationally) for its reliance on terrorism (including suicide bombings), media attention to female combatants tended present the group and its members in a more positive, sympathetic light. Thus, the leadership's willingness to allow journalists to interview female combatants is likely indicative of its awareness of the potential benefits associated with raising public awareness of women's participation in the group.

The LTTE also sought to draw international attention to women's participation through the production and distribution of propaganda films (Brun 2005, 66). In a similar manner, in the early 1990s the LTTE International Secretariat in London published Adele Ann Wilby's (1993) book *Women Fighters of Liberation Tigers*, which provided a vivid illustration of women's participation in the LTTE.[22] As the leader of the LTTE's women's wing, the Australian-born Wilby offered a somewhat embellished and glorified account of the group and its female cadre. However, this is expected, given that in all likelihood the intention of the publication was to cast the group in a more positive light. The book should therefore be interpreted as a work of LTTE propaganda at least as much as an academic resource (Schalk 1994, 176). In the mid-2000s, the leadership also permitted the filming of *My Daughter the Terrorist*, an internationally distributed documentary by filmmaker Beate Arnestad that focused on the experiences of two young female members of the Black Tigers.[23] While arguably more objective than Wilby's account, the film nonetheless humanizes the group's female participants and ultimately casts them as sympathetic figures by explaining the background of the conflict, highlighting how the women think about their decision to join the group, and interviewing their family members. In this sense, it exhibits many of the frames employed by the media that I discussed in chapter 2.

These examples, as well as a wealth of previous research, suggest that the LTTE routinely utilized the presence of female combatants in an attempt to cultivate a more sympathetic and positive image for the group and to generate support for its cause. As with the PKK case, it is difficult to draw a causal connection between the use of female fighters in rebel propaganda and the ability of the group to secure external support. However, there is some evidence that the presence of female combatants provided strategic benefits to the LTTE, specifically its ability to sustain or expand its troop size and its ability to secure (or increase existing) support from a transnational constituency.

At minimum, the expansion of female recruitment almost certainly allowed the LTTE to field more troops than it otherwise could have, thus promoting its survival and its ability to challenge the regime. Heavy fighting in 1990 and 1991 severely depleted the LTTE's forces; nonetheless, it managed to maintain fairly consistent troop numbers over this period of time, suggesting that it was able to rapidly replace the losses sustained during this period (Hellmann-Rajanayagam 1994, 40–41). Other sources suggest that the size of the LTTE's combat force increased in the early to mid-1990s (UCDP 2018). Given that the group apparently increased the number of women in its ranks during this window of time, its ability to maintain (or perhaps expand) its combat force indicates a level of success for the leadership's gambit. Indeed, one source reported that in 1991 more than three thousand women were fighting with the LTTE, representing a substantial increase over the number of female troops present in the mid-1980s (Schalk 1994, 166). The rate of reported women's casualties also increased over this period, rising from only a handful a year in the mid-1980s to 79 in 1990 and to more than 200 in 1991 (Schalk 1994, 166). This suggests that women were taking on more of the burden of fighting and subsequently paying higher costs. Assuming women's participation continued at a similar rate for the rest of the conflict, the number of female fighters in the 1990s allowed the LTTE to field a more durable force capable of presenting a more formidable challenge to the Sri Lankan military.[24]

The increased commitment of financial resources by the diaspora community also corresponds to the period during which women's roles rapidly expanded and the group began utilizing them in its propaganda. The loss of Indian support in the late 1980s pressured the group to locate alternative resource streams. In response, during the early 1990s the LTTE sought to expand the funding it received from the Tamil diaspora in Europe and North America (Högbladh, Pettersson, and Themnér 2011; Staniland 2014, 166).[25] This strategy appeared to be successful; since that time period, the financial resources flowing from the Tamil diaspora community to the LTTE have arguably represented the group's most important source of external support. Indeed, some estimates suggest that the LTTE's annual revenues from diaspora fund-raising exceeded $80 million annually during some years of the conflict (Byman et al. 2001; Staniland 2014, 177; Wayland 2004). Owing to its heavy reliance on diaspora support, the LTTE devoted substantial efforts to fundraising among this community, including engaging in coercion of Tamil business owners. Nevertheless, propaganda consistently played a central role in soliciting donations from Tamils abroad.

As noted above, Tamil propaganda featuring female combatants was intended to shape the perceptions of audiences abroad as well as at home.

This included using the symbol of the female fighters as part of their effort to sustain financial support from the Tamil diaspora (Brun 2005, 66–67). As I discussed in chapter 2, rebel groups often seek to achieve support from abroad by exploiting feelings of guilt and shame among members of the community living in relative comfort overseas while their co-ethnics sacrifice and suffer in the homeland (e.g., Adamson 2013; Wayland 2004). Highlighting women's contributions to the war represents a particularly effective strategy for inducing a sense of guilt among members of the diaspora, particularly male members. As an example, a 1991 issue of *Cutantirap Paravai* featured a letter from a young female soldier to her brother living in Canada. In the letter, the woman pointedly rejects her brother's request for her to emigrate and asserts that Tamil women must suffer and sacrifice because of his cowardice (Hellmann-Rajanayagam 1994, 70).

The audience to which the piece was directed was not explicitly identified. However, the context of the publication suggests that it may have been intended to induce a sense of guilt among members of the diaspora community into supporting the group by asserting that while they were living comfortably abroad, women (sisters and daughters) were fighting and dying to defend and liberate the Tamil community in Sri Lanka.[26] The tactic used in the letter therefore appears quite similar to rebel groups' well-established use of the presence of female fighters as a way to shame reluctant men into joining the rebellion. In both cases, the assertion that a woman would be willing to risk her life while a man sits on the sidelines (sometimes in another country) is intended to directly challenge the target's masculinity and call their courage into question. For men in the diaspora, joining the rebellion may not be a viable option; however, committing some portion of their (relative) wealth to the cause represents one way in which they can attempt to assuage some of the shame and guilt they feel for allowing women to fight the war on their behalf.[27]

Establishing a causal linkage between the visible presence of women in the LTTE and the group's ability to secure or sustain domestic and/or international support is challenging. Nonetheless, there is ample evidence demonstrating that the LTTE frequently relied on the symbolism of its female fighters in their domestic and international propaganda efforts. It is possible, however, that the growth in the prevalence of female fighters and the expansion of LTTE's diaspora fundraising success are both products of the same unobserved process. Paul Staniland (2014, 166), for example, asserts that the group's ability to raise funds from the diaspora community was a function of its organizational structure. If the organizational structure of the group also accounted for the increase of female combatants, then the covariance in the proportion of female combatants and diaspora fundraising may be epiphenomenal.

There is some evidence that the organizational structure of the LTTE evolved during the late 1980s and early 1990s as it struggled to manage resource strains and attempted to govern the areas over which it exerted nominal control (Mampilly 2011, 108–111; Staniland 2014, 169). Yet, the large-scale inclusion of female combatants most likely represented a strategic dimension of this process of organizational change rather than simply a by-product of it. While it would be incorrect to assert that diaspora support was in any way caused by the expansion of female combatants in the LTTE ranks, it is reasonable to believe that their increasing presence and the frequency with which the LTTE utilized female fighters in its propaganda efforts assisted the group in expanding such support and maintaining it over the subsequent decades. In other words, female fighters likely enhanced the group's ability to generate such resources even if they were not directly responsible for them.

Factions in the Zimbabwean War of Liberation

By some estimates, women constituted upwards of 30 percent of the rebel fighting forces of ZANLA and ZIPRA, the respective guerrilla armies associated with ZANU and ZAPU. However, most scholars assert that this figure is significantly overinflated and likely includes all women who supported the movements, including those that had no formal affiliation with the guerilla forces (Geisler 1995, 551; Kriger 1992, 191; O'Gorman 2011, 57).[28] More realistic figures put the proportion of female combatants in the movements' armed wings at closer to 10 percent (Geisler 2004, 51; Lyons 2004, 159–160).[29] Regardless of the specific numbers, by the end of the conflict and the termination of white minority rule in 1979, thousands of female combatants had fought in the Zimbabwean War of Liberation.

While both factions included female fighters and appear to have recruited them at similar levels, the two groups differed markedly in terms of the manner in which they organized and deployed female guerrillas.[30] ZIPRA organized its female combatants into a separate women's battalion, which was rarely deployed outside of the group's camps on the Zambia-Rhodesian border, thereby significantly limiting women's exposure to combat (Alexander and McGregor 2004, 88; O'Gorman 2011, 57). Nonetheless, members of the women's battalion received military training, and some were armed, which occasionally led them to participate in combat. This most commonly occurred when Rhodesian military forces attacked the camps (Alexander and McGregor 2004, 91). By contrast, ZANLA's leadership chose to incorporate its female combatants into existing battalions; as a result, its female guerrillas were more frequently engaged in direct confrontation with Rhodesian forces, especially in the final years of the conflict. While ZANLA typically deployed its female fighters in combat support roles (carrying weapons and ammunition, evacuating wounded male soldiers, conducting reconnaissance, etc.), the nature of the guerrilla conflict often led them into armed encounters with Rhodesia troops (Geisler 2004; O'Gorman 2011; Chris Johnson 1992, 160–162).[31] Consequently, it is clear that female guerrillas directly participated in combat during the Zimbabwean Liberation War.

As in many other rebel movements, the incorporation of women into the groups' combat forces occurred only over time. At the outset of the conflict, only men participated in combat, while women were deployed only in political, logistical, medical, and other support roles and were denied combat training and weapons (Gonzalez-Perez 2008, 82; Lyons 2004, 157). Leda Stott (1990, 27) notes that when women first began arriving in the guerrilla camps, they primarily performed domestic roles such as washing clothes, preparing meals, and tending to the needs of male guerrillas. Yet, by the early 1970s, the roles women played in ZIPRA and ZANLA had begun to change dramatically, and women increasingly received military training and took on military roles (Lyons 2004, 109–110; O'Gorman 2011, 57; Stott 1990, 27–28). By the end of the conflict, thousands of women were incorporated into the rebels' fighting forces, many were participating in combat, and a few had attained leadership positions (Chung 2016; Lyons 2004, 105, 157; Thompson 1982, 247). As with the other cases discussed in this chapter, the change in women's participation in the armed Zimbabwean liberation movements through the 1960s and 1970s appears to have resulted from the combination of the groups' ideological orientations as well as the resource pressures they encountered as the war expanded and its timeline extended.

Ideology and Reluctant Egalitarianism

Both ZANLA and ZIPRA were supported by communist powers, the former by the Soviet Union and the latter by China. Unsurprisingly, this support also mirrored the ideological orientations outwardly projected by the groups' leaderships. Interestingly, however, although both groups had adopted broadly Marxist-Leninist political ideologies, by the early 1970s both employed decidedly Maoist ideologies of guerrilla warfare (Kriger 1992, 89–91). Despite the Marxist rhetoric employed by both movements, numerous scholars have questioned the leaders' commitment to Marxist thought. Instead, these scholars assert that while Marxist ideology influenced some of ZANU and ZAPU's leaders and the groups employed some elements of socialist rhetoric in their propaganda, Marxist ideas were always "tenuous strands" in their ideologies and typically took a backseat to the goal of national liberation (see Kriger 1992, 95–96).[32]

While neither group's leadership was deeply immersed in Marxist political ideology, both groups nonetheless adopted some of Marxism's revolutionary rhetoric with respect to gender equality and inclusion. This was particularly true in the case of ZANLA, whose leadership believed from early on that women's participation was important for the success of the rebellion. Influenced heavily by FRELIMO in Mozambique, whose leadership had insisted that women's liberation was essential for the revolution, ZANU incorporated elements of gender egalitarianism and the advancement of women's rights into the party's official political platform (Stott 1989, 20; Lyons 2004, 121, 176–179). High-ranking ZANU officials such as Robert and Sally Mugabe and Naomi Nhiwatiwa frequently asserted the movement's commitment to gender equality, especially to international audiences, and highlighted (though often exaggerated) the proportion of women in ZANLA as evidence of that commitment (Kriger 1992, 191–192). Moreover, there is some evidence that, like the PKK and LTTE, ZANLA implemented policies that modestly improved women's rights in the areas that it controlled, including the elimination of the dowry, improving women's access to education, and severely punishing men who were accused of beating their wives (Kriger 1992, 192–194).

However, other observers, including some former members of the organization, have asserted that the group's concern for advancing gender equality was largely superficial and motivated more by instrumental than normative interests (Nhongo-Simbanegavi 2000).[33] This is reflected in the group's willingness to vary its attention to promoting gender equality depending on its assessment of the costs and benefits of doing so. While advocating gender equality was consistent with the group's ideology and proved useful at garnering support from the female population and Western audiences, it also provoked antipathy from many segments of the male population. For instance, many male recruits resented having to take orders from higher-ranking female soldiers, refused to use contraception when engaging in sexual relationships with female recruits, and often expressed concerns that women were poorly suited to combat. Similarly, ZANLA's efforts to advance women's rights led to a backlash among men in the communities it controlled (Kriger 1992, 193–196). Consequently, despite initial efforts to address gender equality, ZANLA's concerns about alienating the communities on which it depended for support curtailed many of its stated goals of women's liberation.

The discussion above highlights that the groups were highly sensitive to the strategic costs associated with the pursuit of gender egalitarian policies or other challenges to the traditional norms. Moreover, ZANU's willingness to permit traditional institutions (e.g., the chieftain system) to retain local authority after independence suggests that rebel leaders placed some value on the preservation of culture and tradition (Kriger 1992, 233–234). At the same time, however, the party also adopted numerous laws in the aftermath of the conflict that sought to improve women's status and rights in society, including those giving women majority status at the age of 18, prohibiting gender-based employment and wage discrimination, granting the right to a share property after a divorce, and providing maternity leave (Kriger 1992, 235). Thus, while the leaderships of ZANLA and ZIPRA were not committed Marxists, neither did they actively seek to reestablish the precolonial social order or reaffirm traditional gender norms or gender-based divisions of labor. If anything, the groups' attitudes on women's rights and gender egalitarianism appeared to be highly flexible and contingent on the strategic context.

Resource Demands and Strategic Adaptation

The rebel leaderships' utilitarian approach to gender issues is also manifest in their decisions to recruit female combatants. In this sense, the groups' behaviors are very much in line with the arguments presented in chapter 1. Provided that a group's leadership does not view gender egalitarianism and the erosion of traditional gender norms as inherently antithetical to its political ideology, it should become increasingly willing to recruit female combatants when it experiences acute resource pressures, such as those created by the expansion of escalation of the conflict. In the cases of both ZANLA and ZIPRA, the rate of women's recruitment and the willingness to deploy female fighters in combat roles increased in the later years in the conflict, particularly as it rapidly intensified in the mid-1970s.

Even though women had occupied noncombat roles in both movements since the outset of the conflict in the early 1960s, they only began to receive military training and carry weapons in 1973 or 1974 (Stott 1989, 27–28; Lyons 2004, 107–110). By 1977, however, thousands of female combatants had either been organized into female-only units within ZIPRA or integrated into existing combat units within ZANLA (O'Gorman 2011, 57; Lyons 2004, 157; Gonzalez-Perez 2008, 82). Moreover, it was at this time that ZANLA's leadership first chose to officially deploy women in armed combat, representing a substantial revision of the previous policy (Nhongo-Simbanegavi 2000, 56–57).[34] Fay Chung (2016), a senior political official in ZANU during the war, contends that Josiah Tongogara, ZANLA's commander in Rhodesia, first recognized the strategic value of female combatants and began to encourage the recruitment of women and inclusion of women in all aspects of the rebellion, including in positions of military and political leadership.

As illustrated in figure 3.3, the timing of female recruitment corresponds to periods of increasing conflict intensity. The dotted vertical lines that cross the x-axis at 1974 and 1977 represent the window of time during which most sources suggest that ZANLA and ZIPRA began to increase the number of female combatants within their ranks, provide them with training, and (particularly for ZANLA) deploy them in combat roles. For the first several years, the conflict was largely characterized as low intensity, producing only a few dozen to just over a hundred battle-related deaths per year. By the mid-1970s, however, that figure had grown to several hundred annually and to several thousand per year by the late-1970s. In comparison with the intensity of the conflict during its final years, the increase in violence in 1973 appears fairly modest. However, compared to the previous years of the conflict it was substantial, increasing by a factor of ten between 1970 and 1973. Between 1974 and 1977, however, the rate of annual

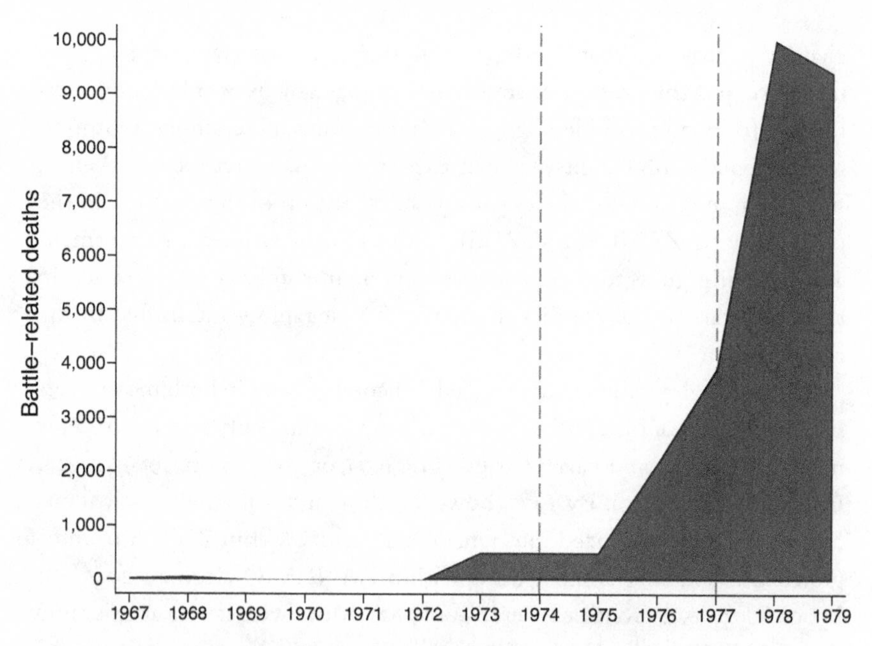

Figure 3.3 Annual battle deaths during the Zimbabwean War of Liberation
Sources: Data on battle deaths from Lacina and Gleditsch (2005) and UCDP (2018)

killings increased by a factor of eight, rising to more than four thousand per year. It was during this period that the conflict escalated to a full-scale civil war, requiring the large-scale mobilization of both human and material resources. And it is during this period that we observe a substantial increase in the number of female fighters.

In addition to the correlation between rising conflict intensity and the increasing rate of women's participation in the rebel movements, anecdotal evidence suggests that rising resource demands played a central role in the decision of the leadership of both groups to recruit larger numbers of women. By 1972 women were viewed as an essential part of guerilla strategy, and by 1973 they were being trained to fight (Lyons 2004, 107, 157). Chung (2006, 81) observes that this period not only saw an intense escalation of the conflict, but also the entry of women as major players in the liberation struggle. As the shortage of male recruits grew more severe, ZANLA began to encourage women to assume a greater role in the rebellion and began to train them as fighters (Stott 1989, 27–28). Several scholars contend that the realities of the armed struggle, rather than any commitment to social transformation, pressured the leadership to admit women into the rebellion alongside men (Geisler 2004, 50; O'Gorman 2011, 57). While ZANLA's leadership had initially taken the position that women were not suited to the rigors of guerrilla warfare, they increasingly came to believe that they could not prevail in their liberation struggle without broadening the revolutionary base and recruiting women into the movement (Geisler 2004, 50; Stott 1990, 27). Indeed, Chung (2016) asserts that this period represented a turning point in the war, and, without the participation of female fighters, success would have been much more difficult to achieve.

Propaganda, Recruitment, and External Support

While ZANLA and ZIPRA did not utilize propaganda or engage the international media with the same level of sophistication as the PKK or the LTTE, both groups devoted substantial efforts to garnering domestic and international support for the rebellion. In addition to their efforts to mobilize support among the African population in Rhodesia and to secure or maintain support from the governments of bordering states (principally Zambia and Mozambique), they also attempted to garner attention and support from sympathetic anticolonial movements in the West as well as the governments of both socialist and Western democratic states. In many cases, particularly in the West, the dismantling of white settler rule in Rhodesia became subsumed under a broader anti-apartheid movement that sought to liberate all of Namibia and South Africa as well.

Like the other rebel movements discussed in this chapter, the image and presence of female combatants played an important role in the group's propaganda and recruitment efforts. As Eleanor O'Gorman (2011, 57) observes, ZAPU and ZANU leaders eventually recognized the effectiveness of female fighters as a propaganda tool. As such, as women's participation increased, both groups included images of female guerrillas in their propaganda materials, including those oriented toward international observers (Lyons 2004, 158). Critical evaluations of the ZANU/ZANLA leadership's commitment to women's rights and gender egalitarianism have often asserted that the movement's efforts to highlight women's participation in the rebellion, particularly as combatants and in leadership positions, amounted to little more than propaganda directed at external audiences (e.g., Nhongo-Simbanegavi 2000). These critics assert that in reality female guerrillas had to challenge a reluctant male leadership in order to gain a measure of equality with their male counterparts and to be taken seriously as soldiers. Yet, the leadership often highlighted their presence and used the very few women in leadership positions to advance the narrative that the movement was both progressive and had broad social support.

Promoting an image of legitimacy and a commitment to liberal ideals was crucial for the rebel forces in their effort to secure foreign support, particularly among Western audiences (Mazarire 2017). Eastern Bloc states were generally more willing to extend support to the rebels due to their nominally socialist ideology and their strategic utility in the proxy wars being waged against the West, and other developing states were often inherently supportive of the movements' struggle against colonial domination.[35] Western audiences, however, were more ambivalent about supporting such anticolonial liberation movements. Even if Western governments and citizens were not supportive of the white minority government and often condemned their abusive policies and intransigence, they were not necessarily inclined to support armed movements that received assistance from the Soviet Union and China.

Highlighting the presence of female combatants in the ranks represented one way that ZANLA and ZIPRA attempted to appeal to Western audiences, particularly liberal and left-wing political movements and activist organizations in the United States and Europe. While it is difficult to assess the effectiveness of this strategy, it is worth noting that a number of advocacy organizations located in North America and Europe engaged in lobbying, fund-raising, and other forms of activism on behalf of ZANU and ZAPU (Kössler and Melber 2002; Mazarire 2017; Sellström 2002). These included LSM, the African Liberation Support Committee, the Southern African Liberation Committee, and various other groups within the broader anti-apartheid movement. Moreover, as demonstrated by the posters and buttons presented in chapter 2, the image of the female guerrilla became an important symbol in the propaganda materials of these organizations, reflecting the salience of female combatants to those audiences.

Lastly, according to some observers, the presence and contributions of female combatants represented an important aspect of the rebels' domestic resource mobilization and support effort. The role of women in the Zimbabwean Liberation War was no different. Equally important, however, female combatants facilitated the rebels' ability to mobilize support among the population more broadly. The tenacity and courage of the female guerrillas led many Zimbabwean peasants to conclude that if women could risk undertaking such difficult work and endure such sacrifices for the liberation of the country and its people, they too should contribute to the war effort in a meaningful way (Chung 2016). This observation is consistent with the arguments put forth in previous chapters: the image of female combatants often represents an important tool for garnering support and sympathy from the local population.

Conclusion

In this chapter, I have undertaken preliminary investigations of the hypotheses outlined in the previous chapters by examining the dynamics of female recruitment in three specific cases: the Kurdish PKK, the Tamil LTTE, and ZANLA/ZIPRA in Zimbabwe. Each represents an example of a case in which the presence of female combatants is well documented, which allowed me to conduct a more in-depth investigation of the timing of the decision to initiate or expand the recruitment of female combatants. Relatedly, I am able to broadly link the timing of such recruitment to changes in the conflict-related resource demands imposed on the groups. Finally, I am also able to provide greater attention to the ways in which the groups used the presence of female combatants in their propaganda efforts and explore the potential ways in which such efforts influenced the groups' ability to secure and maintain support from domestic and international audiences.

The cases discussed above help illustrate the mechanisms articulated in prior chapters and, as a whole, provide additional evidence for the arguments therein. First, while women were present in both the LTTE and PKK from the outset of the conflict, their prevalence and the roles they played increased substantially over time. In the case of the PKK, women appear to have served in combat units from the beginning, but their participation surged in the 1990s as the conflict entered a particularly intense phase and resource demands increased. For the LTTE, women appear to have occupied largely noncombat roles until the mid-1980s, at which point they began to serve in combat in small numbers. As the resource pressures grew with the loss of Indian support and rising competition with the EPRLF, the LTTE appears to have gradually increased the number of women in its combat forces. That number, however, rapidly increased after about 1990, when the conflict entered a new, more intense phase often referred to as the Eelam War II (1990–1995). In ZANLA and ZIPRA, however, women occupied only support roles until the mid-1970s, when resource pressures and a shortage of male troops pressured the leadership to train, arm, and send women to the front.

These case illustrations also highlight the roles that group ideology played in the evolution of women's participation. As an avowedly Marxist movement, the PKK was minimally constrained by the prevailing social norms of the conservative Kurdish society on which it relied for support. Overturning extant social and political institutions was a core aspect of its ideology, and it therefore was willing to recruit women and deploy them in combat from very early on. Nonetheless, the evidence suggests that the PKK's commitment to women's rights and gender egalitarianism deepened over time, particularly during the period in which women were being recruited at higher numbers. This suggests the likely possibility that the increase in gender egalitarian rhetoric was at least partly instrumental and represented a way for the PKK's leadership to mobilize the larger number of female combatants they needed to sustain them in the conflict.

The LTTE followed a somewhat similar evolution, though the group's adherence to prevailing social values was initially comparatively stronger and the extent of its change was more dramatic. The LTTE did not embrace a Marxist ideology; however, it did adopt some revolutionary beliefs, including espousing greater rights and opportunities for Tamil women. As with the PKK, the group's use of a gender egalitarian rhetoric grew over time and increased as it sought to bring more women into the movement when it faced an increasing demand for troops. Similarly, while lacking any deep commitment to Marxism or Maoism, both ZANLA and ZIPRA espoused a nominally socialist political platform. Both also employed the language of women's rights and emancipation in their rhetoric. However, while they eventually deployed women in combat roles, this decision appeared to come later and more abruptly than in the other movements, and only after a steep escalation in the intensity of the conflict.

Last, the cases highlight the specific ways in which the movements utilized images of female combatants to shape the perceptions and beliefs of domestic and international audiences. While the PKK's leadership appears to have recognized the utility of female fighters in generating additional support from domestic constituencies early on in the conflict, there is little evidence that they explicitly sought to use female combatants to leverage support from international audiences until relatively recently. Since at least the late 2000s, the PKK and sympathetic transnational organizations have intentionally highlighted the high prevalence of female combatants in its ranks to garner attention and sympathy for the group and its cause. By contrast, the LTTE recognized the benefits of highlighting women's participation quite early, producing numerous videos and publications that showcased female cadres' dedication and sacrifices in the conflict. It is not clear precisely when during the conflict ZANLA and ZIPRA began utilizing female combatants in its propaganda efforts. Nonetheless, it appears that by at least the mid-1970s the movements (particularly ZANLA) intentionally highlighted women's roles as fighters and their inclusion (even if modest in reality) in leadership positions to establish its legitimacy and attract external support.

CHAPTER 4

Empirical Evaluation of Female Combatant Prevalence

In chapter 1 I presented a theory of rebel mobilization of female fighters that coupled core insights from existing demand-side theories of rebel recruitment with leader expectations of and tolerance for the potential costs associated with this decision. More specifically, I argued that the leaders of armed groups that face resource demands that outstrip the supply of potential male recruits will become increasingly likely to create opportunities for women to participate in the armed wing of the movement, including allowing them to participate in combat. Yet, I also argued that not all groups facing resource mobilization challenges opt to recruit women; rather, the decision to respond to resource shortages by incorporating women into the movement is contingent on the group's (and its constituents') commitment to traditional gender norms. These arguments produced a series of hypotheses that linked the prevalence of female combatants in an armed group to (1) the severity of conflict-induced resource demands, (2) group political ideology, and (3) the interaction of these factors.

The case illustrations presented in chapter 3 probed the plausibility of the hypotheses and highlighted the dynamics of women's mobilization for war and its implications. The quantitative analyses presented in this (and the subsequent) chapter represent a more systematic analysis of the factors associated with the decision to deploy female combatants. I discuss the data used to evaluate the hypotheses presented in chapter 1 and the results of the quantitative analyses in this chapter. I reserve the examination of the strategic implications of female fighters, which was the topic of chapter 2, for chapter 5.

The theory outlined in chapter 1 assumes that rebel leaders are able to coarsely evaluate the expected tradeoffs involved with adopting a given strategy. Herein, this trade off, which I describe as the dilemma of female recruitment, reflects the balance of the expected benefits of recruiting women (easing resource constraints) and the expected costs associated with that decision (e.g., constituency backlash or internal dissent). Fully assessing the rationale that guides rebel leaders to recruit or exclude women is exceedingly difficult, and it is not my intention, nor is it possible using the data and methods I adopt, to thoroughly investigate the individual motives of any specific leader.[1] Nonetheless, in this chapter I examine a series of indicators that should serve as reasonable proxies for the factors that I argue are likely to influence rebel decision-making. The results of the analyses allow me to identify a set of general conditions that influence the prevalence of female combatants in armed resistance movements.[2]

Concepts and Measures

In order to systematically assess the hypotheses presented above, I conduct a series of statistical analyses using novel cross-national data on the presence and prevalence of female combatants in large sample of contemporary rebel movements. Before delving into the technical aspects of the methodology, it is useful to first identify the variables utilized in the analyses and discuss their ability to capture the core concepts reflected in the prior arguments. In this section I therefore devote substantial attention to the measure of female combatant prevalence used in the analyses, including the operational definition of "female combatants" and the data collection process. I also discuss the relevant predictors and control variables used in the statistical analyses.

The Women in Armed Rebellion Dataset (WARD)

Systematically examining the relationships posited in the hypotheses above requires data on the presence and prevalence of female combatants in a variety of rebel movements. I therefore utilize measures from an updated version of the Women in Armed Rebellion Dataset (WARD) (Wood and Thomas 2017) as the dependent variables in all analyses presented in this chapter. Similar datasets capturing various aspects of women's inclusion in armed groups have emerged in recent years, demonstrating the interest and efforts devoted to understanding patterns of female participation in armed movements. However, each faces shortcomings that limit its usefulness for assessing the hypotheses presented above. For instance, both the Thomas and Bond (2015) and Henshaw (2016a) datasets contain only binary measures reflecting the presence or absence of female combatants. Moreover, the Thomas and Bond (2015) dataset only includes armed groups in Africa while Henshaw's (2016a) and Loken's (2017) datasets are limited to armed groups in the post–Cold War era. Finally, Loken's (2017) dataset is organized at the conflict level, preventing an assessment of the group-level factors that motivate the recruitment of female combatants. I therefore rely on a revised version of WARD for the analyses conducted in this chapter because of its superior geographic and temporal coverage and because it includes an estimate of the prevalence of female fighters within an organization rather than a simple binary measure of their presence.[3]

The updated WARD includes information on the prevalence of female fighters for a sample of more than 250 rebel organizations active between 1964 and 2009.[4] The list of groups included in the UCDP Dyadic Dataset (Harbom, Melander and Wallensteen, 2008) defines the sample of rebel groups included in WARD. Because the primary intention of WARD is to identify female fighters in rebel groups, international conflicts and civil conflicts that involve only military factions (e.g., coups) were excluded from the analysis. Ultimately, WARD includes data on approximately 70 percent of cases included in this subset of the UCDP Dyadic Dataset.

Many of the cases for which data are missing represent examples of factions of other rebel movements. For example, the FMLN was formed as an alliance of five distinct rebel movements,[5] two of which are included as distinct actors in the UCDP Dyadic Dataset along with the FMLN. However, despite the substantial amount of information available on women's participation in the FMLN, very few of the relevant sources located in the data collection process explicitly discuss women's involvement in the different factions. As such, WARD includes only a single score for the umbrella group (FMLN). Similarly, the UCDP dataset identifies a variety of irregular and formal Croatian, Bosnian, and Serb forces involved in the interconnected conflicts in the former Yugoslavia. While there is sufficient information to determine the presence of female fighters in the Serbian Republic of Krajina, Serbian Republic of Bosnia-Herzegovina, and Croatian Republic of Bosnia-Herzegovina, there is insufficient evidence to determine whether or not women participated in the various "irregular forces" that the UCDP includes as participants in the conflict. Similarly, for some groups there is very little information.[6] For instance, coders were unable to locate information of any kind about the composition, attitudes, or actions of the Lahu National United Party (LNUP) in Myanmar or on the Mouvement du Salut National du Tchad (MOSANAT) in Chad, let alone on women's roles in the movements. As such, these groups are coded as missing.

The temporal domain of the dataset was chosen for both practical and theoretical reasons. Data collection is highly time intensive, and information generally becomes less reliable and less readily available as the time frame stretches further into the past. Women's participation and roles within armed groups are not reported in any standardized manner; nor are these data stored in any centralized repository. Gathering such data requires an extensive search of a wide range of resources, including news reports, academic accounts (e.g., books and articles), biographies, governmental sources, and international and NGO reports. While the coding process required a substantial investment of human resources, it was nonetheless important to ensure that the dataset included a wide variety of civil conflicts occurring in many different locations, organized around a range of different political disputes, and operating in different global contexts. Achieving sufficient variation among these dimensions therefore necessitated coding a global sample of groups active during and after the Cold War. Importantly, the variables included in WARD provide only a snapshot of women's participation in armed groups. While the ideal dataset would account for variations in the proportion of female fighters over time, the same information constraints that prevent a more precise measure of female combatant prevalence (see below) prevent the construction of a time-series dataset. Consequently, the dataset only captures cross-sectional variation in women's participation in armed rebellions.

The conceptualization and operationalization of female fighters draws on definitions commonly used in disarmament, demobilization, and reintegration (DDR) programs sponsored by the United Nations and related international organizations. These programs often differentiate *female combatants* from *females associated with armed groups* (UN Women 2012, 22–23).[7] Specifically, they define these categories as follows:

Female combatants: Women and girls who participated in armed conflicts as active combatants using arms.

Female supporters/females associated with armed forces and groups: Women and girls who participated in armed conflicts in supportive roles, whether coerced or voluntarily. These women and girls are economically and socially dependent on the armed force or group for their income and social support. Examples: porters, cooks, nurses, spies, administrators, translators, radio operators, medical assistants, public information workers, camp leaders, or women/girls used for sexual exploitation.

The variables included in WARD reflect these basic definitions. In the dataset, *female combatants* refers to all female members who underwent military training, received combat arms, and participated in organized combat activities on behalf of the organization in any capacity at any time during the conflict. This definition excludes women and girls who exclusively occupied noncombat support roles such as those noted above. It is important to note, however, that while female members of rebel movements often occupy multiple roles that blend combat and noncombat activities, key criteria for the noncombat designation is that these women did not receive combat training, carry weapons, or engage in the direct production of violence.

One of the primary challenges in coding information on women's participation in rebel movements is that the boundary between formal combatant and other roles is often blurred, particularly in the case of guerrilla conflicts. Consequently, the definition of combatant applied by WARD includes women employed in an array of activities ranging from frontline infantry to local militias, to women who were primarily deployed in support roles but who, by virtue of their training and access to weapons, engaged in combat when the situation demanded. For example, Viterna (2013, 129–130) notes that a substantial proportion of female FMLN guerrillas were employed as "expansion agents" whose primary role was recruitment and community engagement. Yet, these (armed) units were occasionally called into battle when frontline forces required additional support and engaged in combat if, in the conduct of their primary duties, they encountered government security forces. Similarly, the majority of ZANLA's female guerrillas were assigned to support roles such as carrying ammunition and weapons to the front and helping to evacuate wounded soldiers from it. While not considered frontline combat troops by the rebel leadership, these women sometimes engaged in combat because their assignments put them in close proximity to enemy troops. Consequently, as in other guerrilla conflicts, women were often on the frontlines of the Rhodesian conflict because there was no clear boundary between the front and the rear (Lyons 2004, 170). More specifically, the operational criteria for *fighter* used in the coding process includes any references to women undertaking any of the following activities on behalf of the rebel group:

- Using arms in combat, including during defensive actions (e.g., protecting camps, returning fire when attacked during noncombat operations, etc.) or against civilian targets
- Operating artillery or antiaircraft weapons against enemy targets
- Service in auxiliary and militia forces, provided that they sometimes participated in offensive or defensive combat operations
- Detonating mines or other explosives against enemy or civilian targets
- Conducting assassinations
- Conducting suicide bombings

Making a determination of the presence of female combatants in an organization required the confirmation of three independent sources. When reports explicitly stated that women did not participate in combat, when women's roles were described as exclusively supportive (e.g., caregivers, fundraisers, couriers, etc.), or when it was not possible to locate any evidence of women participating in combat (despite locating substantial information regarding other group characteristics), the group was coded as not including female combatants. I discuss the challenges of determining the absence of female combatants in more detail below.

The primary variable used in the analyses, *female combatant prevalence*, is a categorical indicator roughly accounting for the estimated proportion of a group's combat force composed of women. The categories range from 1, indicating "no evidence" of female combatants, to 3, representing a "high" prevalence of female combatants. WARD relies on a categorical indicator rather than a direct estimate of the proportion of female combatants in an armed group largely because different sources sometimes provide varying estimates of the numbers of women serving as combatants. For example, different sources report the proportion of women in the LTTE as ranging between 15 percent and 30 percent (e.g., Alison 2004, 450; Stack-O'Connor 2007a, 45). Where available, the coding relies on reported percentages or numbers of female troops.[8] However, in many cases, the sources provide only qualitative descriptions of the extent of women's participation (e.g., "rare," "small numbers," "very few"). This is particularly true of groups that appear to have few female combatants. The blunter coding scheme therefore reflects a tradeoff between precision and confidence in the measure. The dataset also contains the variable *female combatants*, which is simply a dichotomization of the prevalence variable, as well as additional variants of the prevalence measure, which are discussed in more detail below. Table 4.1 briefly describes the categories of *female combatant prevalence.*

Table 4.1 Description of Female Combatant Prevalence Categories

Category	**Prevalence**	**Estimated percentage**	**Examples**
0	No evidence	0	Mouvement des Forces Démocratiques de Casamance (MFDC) Abu Sayyaf Group (ASG) Eritrean Liberation Front (ELF)
1	Low	< 5	Popular Front for the Liberation of Palestine (PFLP) Provisional Irish Republican Army (PIRA) Resistência Nacional Moçambicana (RENAMO)
2	Moderate	5–20	Euskadi Ta Askatasuna (ETA) Liberation Tigers of Tamil Eelam (LTTE) Contras
3	High	> 20	Movement for Democracy in Liberia (MODEL) Fuerzas Armadas Revolucionarias de Colombia (FARC) Tigrayan People's Liberation Front (TPLF)

Note: All examples reflect the "best" estimate of *female combatant prevalence* from WARD.

Owing to both conflicting estimates of the extent of women's participation and the fact that available sources do not always clearly differentiate between female combatants and women who engaged in noncombat roles, WARD also includes an alternative "high" estimate of the female combatant prevalence indicator. The case of GAM in Aceh is illustrative. While some available sources suggest that women may have participated in combat operations on rare occasions (Aspinall 2009, 93; Jaulola 2013, 36–37), others assert that women were explicitly denied combat roles or strictly limited to support roles (Barter 2015, 349; Schulze 2003, 255). Based on the overall weight of available evidence, GAM is assigned to category 0 in the more conservative "best" estimate variable. However, it receives the alternative score of 1 in the less restrictive "high" measure of female combatant prevalence.[9]

Similarly, as I discussed in chapter 1, suicide bombers represents a distinct class of combatants, and female suicide bombers are arguably unique in that in some armed groups they represent the only "combat" role allowed to women. In such cases, they typically represent only a tiny fraction of the group's combat force, often numbering in the single digits over the span of the entire conflict. Moreover, they seldom appear to receive the same training or are given the same responsibilities as other fighters. Considering female suicide bombers as "combatants" may therefore amount to conceptual stretching because of the unique nature of the role. Consequently, WARD also includes an additional version of the *female combatant prevalence* variable that excludes cases in which women were exclusively used as suicide bombers and were otherwise denied combat roles in the organization.

A persistent challenge in coding WARD is accurately assessing the absence of female combatants. Indeed, the possibility of false negatives, in which a group is potentially incorrectly coded as having no female fighters when in fact it included some number of female fighters, represents a much larger challenge than false positives, where a group is incorrectly coded as having female fighters when no women ever participated in combat. While there is often variation in the details regarding the number of women in the organization and how frequently they participated in combat, there are few cases where there are competing assertions about women's participation in conflict. WARD includes twenty-five cases (less than 10 percent of the sample) for which there is some ambiguity as to whether or not women ever participated in combat. In the vast majority of these cases, the different "best" and "high" scores assigned to the group reflect the use of vague language regarding women's roles in the source documents. For instance, there may be references to "female cadres," "women rebels," or "women taking up arms," which obscure the particular role women played in the group and make it difficult to discern whether or not these female members of the group ever participated in combat. The "best" measure therefore reflects the conservative case that these women were members of the group but not combatants, while the "high" score reflects the more liberal interpretation of the terms and considers these women to be combatants. Thus, unless multiple sources unanimously inaccurately report that women served in combat when they in fact did not, which seems unlikely, the risk of false positives appears low.

The absence of information regarding women's roles in the rebel organization presents a greater challenge. In these cases, establishing with any degree of certainty that women did not serve in combat roles is difficult. As noted above, the sample included in WARD already excludes a number of rebel organizations for which a dearth of information exists and for which no information on women's participation was located. For the remaining 250-plus groups, coders were able to locate substantial information about the group, including information about its origins, leaders, activities, ideology, and so on. In most cases, coders also located information explicitly discussing the roles of women within the organization, women's support for the group, and/or detailed discussions of the group's gender attitudes. In some cases, there was clear information that women were members of the group's political or military wing and actively supported the group's military operations but did not participate in combat. For instance, one source on women's roles in the MILF in the Philippines asserts that while women are not allowed to serve as combatants, women do assist the movement with "medical, communication, and other auxiliary needs" through its women's wing (Santos and Santos 2010, 359). Similarly, a source used to code the MFDC in Senegal contends that "although there have been no reports of female combatants," women assist the rebellion by providing food, ammunition, transportation and other services (Stam 2009, 343).

Reports explicitly noting the absence of women in the organization or describing their roles within the movement were located in roughly 71 percent of the cases included in WARD. The remaining 29 percent represent the subset of cases for which coders were able to locate substantial information about the group but were unable to locate clear references to women's roles within the group, information confirming or denying the absence of women combatants, or any relevant information about the group from which women's participation could be inferred. The absence of female combatants is assumed in these cases. Given that women's participation in combat—or even their participation in noncombat military roles—is so frequently viewed as a novelty or as a challenge to existing societal norms, it seems unlikely that the presence of women would go completely unreported. Nonetheless, WARD includes a binary variable entitled *low information* to denote these cases. As robustness checks for the analyses presented below, I rerun the models on only the subset of cases for which explicit information on female combatants was located.[10]

The distributions of the versions of the various measures of *female combatant prevalence* for the sample are presented in figure 4.1. The histogram located in the left-hand panel of the figure shows the distribution for the variable representing the best estimate of female combatant prevalence. According to that estimate, almost 61 percent of the groups in the sample show no evidence of female combatants, while roughly 22 percent have a low prevalence of female combatants, and 11 percent and 6 percent, respectively, have a moderate prevalence and a high prevalence. The trend is similar across the other two variables accounting for female combat prevalence, though the distributions change slightly. The less restrictive high estimate of the measure, which is illustrated in the center panel of the

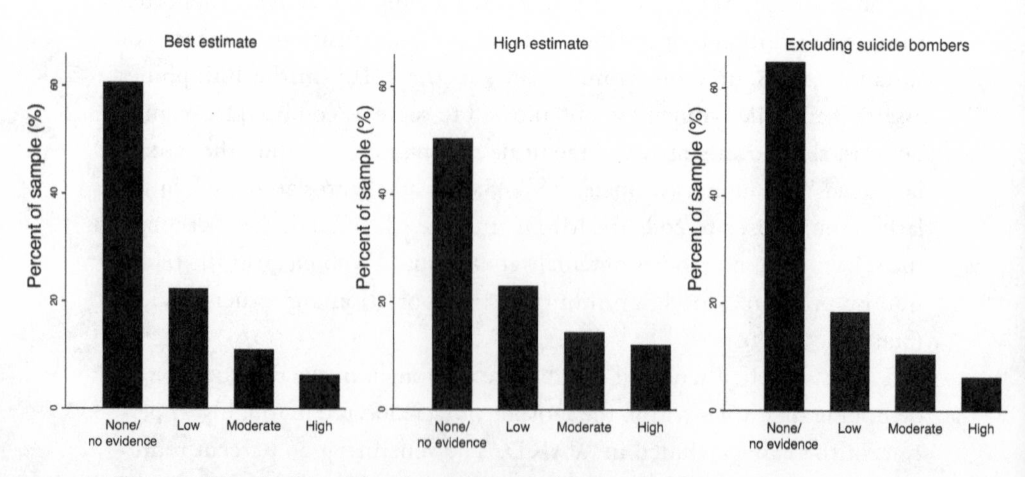

Figure 4.1 Distribution of female combat prevalence categories

figure, reduces the number of groups in the sample without evidence of female combatants to approximately 51 percent, slightly increases the proportion of groups coded as having a moderate prevalence of female combatants, and almost doubles the number of groups reported as having a high level of female combatant prevalence.

The right-hand panel shows the distribution of cases in the sample using the version of the measure that excludes cases in which women only participated as suicide bombers. This measure results in a reduction in the proportion of groups coded as having a low prevalence of female combatants and a corresponding increase in the groups that were coded as having no female combatants, but otherwise the distribution is unchanged. This is not surprising since suicide bombers rarely account for more than a handful of the fighters in any group, and female suicide bombers represent a small subset of all suicide bombers in most cases. Overall, there are eleven groups included in the sample (about 4 percent of the total) in which women served as suicide bombers but were excluded from any other combat roles. Despite these differences, the measures paint very similar overall pictures of the prevalence of female combatants in rebel groups.

Figure 4.2 shows the temporal variation of the "best" estimate measure of *female combatant prevalence*.[11] The distributions of the data across the four categories of the variable are fairly similar, which suggests that the prevalence of female combatants has not radically changed over the decades for which data are available. However, there is one trend in the data that deserves some additional scrutiny. Across the decades represented in the panels, it appears that the number of cases for which there is "no evidence" of female combatants declines as the temporal window advances, while the number of cases in which there is evidence of a "low" prevalence of female fighters increases. The change is not particularly dramatic—for instance, the proportion of cases for which there is "no evidence" of female combatants declines by 7 percent between the first and second panels and by 4 percent between the second and third panels. Nonetheless, the trend may be indicative of greater information availability for more recent conflicts.

Figure 4.3 shows the geographic variation in the prevalence of female fighters. Specifically, the figure shows the highest level of the "best" estimate measure of *female combatant prevalence* for all groups active in the state between 1964 and 2009. States that did not experience a civil conflict during this time period are colored white. States that experienced a civil conflict but for which there is no evidence of female fighters within the rebel

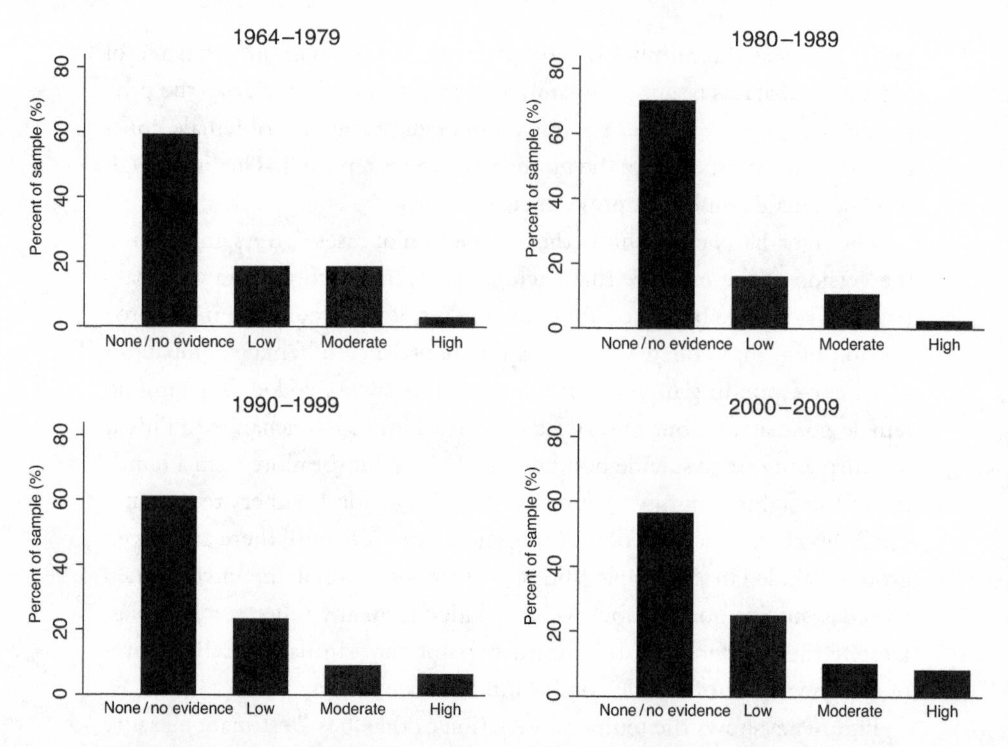

Figure 4.2 Distribution of female combat prevalence categories by time period

group(s) that fought in the country during the specified time period are indicated with light gray shading. Medium gray shading indicates states in which at least one group had a "low" prevalence of female fighters. Dark gray reflects a "moderate" prevalence of female fighters, and black indicates a "high" prevalence of female fighters in at least one group in the state.

The figure confirms previous assertions that female fighters are a global phenomenon (e.g., Mazurana et al. 2002). Indeed, some number of female combatants were present in roughly three quarters of countries experiencing a civil conflict during the period specified. However, the figure also illustrates some geographic variation. Particularly, female combatants are most prevalent in Latin American conflicts during this period. In addition, the high frequency of female fighters across Latin America and within a number of contiguous countries in southern and western Africa and the absence of female combatants in North Africa suggests a certain degree of geographical clustering.[12] It is important to note, however, that many

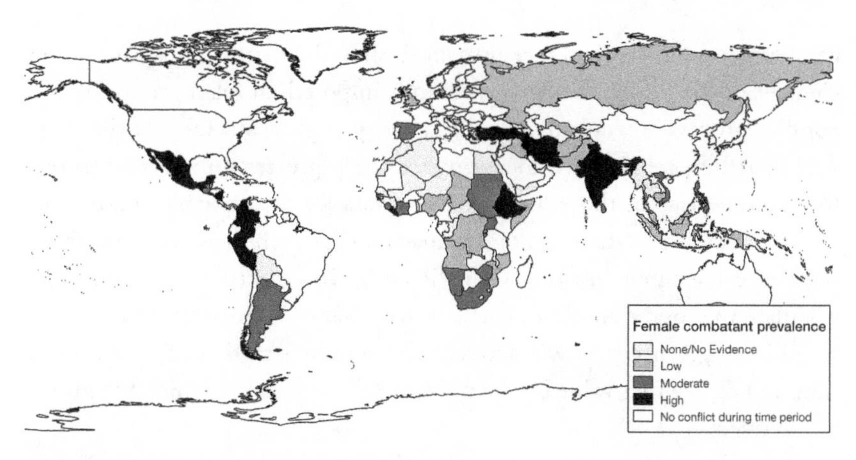

Figure 4.3 Maximum female combatant prevalence by country, 1964–2009

countries hosted multiple different rebel groups during this time period. Given that most rebel groups exclude women from combat roles, the variation in the presence of female combatants seems to be greater across groups than across countries. This suggests the need to consider group-level characteristics when theorizing about the factors associated with the presence and prevalence of female fighters.

Resource Demands and Conflict Costs

In chapter 1, I hypothesized that as the severity of the conflict increases, the prevalence of female combatants is expected to increase. Ideally, it would be possible to account for the specific resource constraints imposed on rebels during the conflict. For example, an estimate of the ratio of recruitment rates (supply) and loss rates (demand) would effectively capture the recruitment shortfalls in human resources that I hypothesize drive rebel leaders' decisions to open combat positions to women. Unfortunately, the components necessary to construct this measure do not readily exist. Consequently, I rely on the annual rate of battle-related deaths as a proxy measure for resource demands.

The variable *annual battle deaths* represents the natural log of the average number of deaths per annum that are attributed to the conflict in which the group was engaged. While this variable does not explicitly capture the gap between and loss and recruitment, it should nonetheless capture some element of the human resource demands imposed on rebel groups by the conflict process. Specifically, as the severity (e.g., death rate) of the conflict increases, rebels must locate increasingly greater numbers of troops to take the place of those lost in battle. Data for this variable come from the Battle Deaths Dataset (v3.0) (Lacina and Gleditsch 2005) available from the Peace Research Institute Oslo (PRIO). To construct this variable, I calculate the total number of battle deaths that occurred during the years in which a given group was actively engaged in armed conflict with the state and divide that value by the number of years during which the group was active.[13]

One potential issue associated with the use of the data from the Battle Deaths Dataset is that it reflects the total number of deaths from a conflict—including those accruing to the state and to any other nonstate actors involved in the conflict—rather than the number of deaths suffered by a specific rebel movement. For this reason, I also employ a similar measure created using data from the UCDP Battle-Related Deaths Dataset (Allansson, Melander, and Themner 2016) as a robustness check. This measure captures the number of deaths suffered only by the specific rebel group under observation. The primary drawback of this measure is that its temporal domain extends back only to 1989, nearly halving the sample size. Nonetheless, the results are generally robust to this measure.

Political Ideology

In order to evaluate the hypotheses related to group ideology, I rely on the binary indicators previously constructed by Wood and Thomas (2017).[14] To construct their measures, the authors relied on multiple existing databases, including the Nonstate Armed Groups (NAGs) Dataset (San-Akca, 2015); the (now-defunct) Terrorist Organization Profiles (TOPs), which were previously available from the National Consortium for the Study of Terrorism and Responses to Terrorism (START 2016), and the Big Allied and Dangerous (BAAD) Dataset (Asal and Rethemeyer 2015). Overall, the ideologies assigned to the groups in the sample are highly consistent across the various sources. However, in the few cases in which the coding differs, additional sources were consulted to make a determination. These databases typically assign groups to a few broad ideological categories, including leftist, rightist, religious, nationalist, no ideology, and so on. Using this information, Wood and Thomas constructed the following mutually exclusive binary indicators reflecting a group's primary ideology: *leftist*, *religious*, *secular nationalist*, and *secular (other)*. While these categories are not exhaustive, they reflect the primary ideological categories of the majority of the rebel groups included in the sample.

Leftist rebellions include all groups that adopt a Marxist-inspired ideology (e.g., socialist, communist, Maoist, Marxist-Leninist, and so on). *Religious* groups are those that organize specially around a defined religious doctrine or set of religious beliefs. In most cases, these are groups who seek to impose their group's religious doctrine on the areas and/or population over which they exercise control or seek to implement a legal or political system based explicitly on their group's religious principles. Importantly, having a membership drawn primarily from one religious community or simply organizing around a specific religious identity is not sufficient to be coded as having a religious ideology. Such groups are generally coded as nationalist. Rather, religious movements must advance a set of social or political objectives that reflect the belief system or values of their specific religious community and seek to impose them on the areas they govern or control.

Secular nationalist groups mobilize on behalf of a distinct ethnic or national community with the intention of advancing the interests of that community, often through establishing autonomy or self-governance. I exclude nationalist movements that also advance religious or Marxist goals from this category in order to isolate the specific effects of each ideology. However, in alternative tests discussed below, I further disaggregate leftist and religious groups by those that also include a nationalist component in order to assess the extent to which nationalist beliefs interact with the other ideological categories. Finally, the variable *secular (other)* includes all remaining groups that did not advocate a nationalist, religious, or leftist ideology. This includes groups that did not appear to espouse a specific ideology as well as a handful of groups that espoused other ideologies that did not fit in the other categories.[15]

The distribution of ideologies for the rebel groups included in the sample is presented in figure 4.4. In the sample, 20 percent of the groups espoused a Marxist, Maoist, or other leftist ideology, while 21 percent of groups adopted a religious ideology. Thirty-four percent of groups in the sample adopted a secular nationalist ideology that did not include any

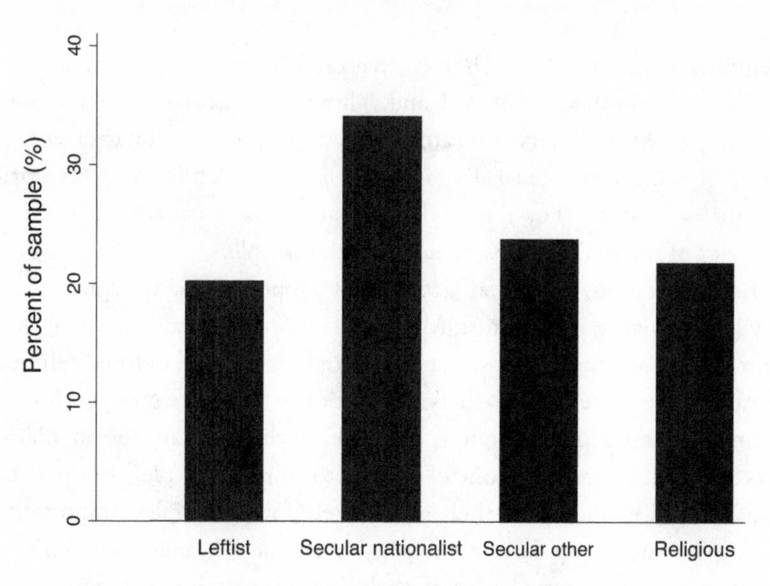

Figure 4.4 Distribution of ideologies for rebel groups in the sample

variant of Marxism, and 24 percent of groups adopted some other secular ideology (again excluding leftist ideologies). As noted above, the ideology variables are constructed such that nationalism is not a mutually exclusive ideology. Rather, with the exception of *secular nationalist*, nationalist groups that also adopt Marxist or religious ideologies are subsumed by those ideological categories. Overall, 47 percent of the groups included in the sample espouse some form of nationalist ideology or advance nationalist goals. Roughly 71 percent of all nationalist groups are included in the category *secular nationalist*, while 15 percent are included among the *leftist* category and 14 percent are included in the *religious* category.

Other Independent Variables

I also include several potentially relevant confounding variables in the statistical analyses. These variables are intended to account for alternative factors that may influence the presence or prevalence of female combatants within a rebel movement. The first such variable is the duration of the conflict. Anecdotal evidence suggests that rebel groups are often initially reluctant to recruit female fighters but eventually become more willing to include women in combat roles. As I argued, resource demands, which often become more acute over time, are a primary driver of the decision to incorporate female combatants into the movement. However, it is possible that duration exerts an independent influence on leaders' decisions, such that over time groups simply become more willing to accept women into the ranks and to deploy them in combat. This could occur either because women eventually pressure the group to include them and to provide them with more expansive, prestigious roles or because over time women are able to effectively demonstrate their capabilities and commitment to skeptical rebel leaders. For this reason, I include the variable *duration*, which is operationalized as the log-transformed value of the number of years between the group's initiation of violent conflict with a government and its cessation of those hostilities as reported in the UCDP Dyadic Dataset.[16]

I also include a variable accounting for the use of forcible recruitment strategies by rebel groups. I include this variable because rebel groups that rely on them are often indiscriminate in their selection and may be more likely to recruit female fighters to fill resource needs. Groups that face acute resource demands are also more likely to engage in forced recruitment, which necessitates the inclusion of a variable accounting for this recruitment strategy. The binary measure *forced recruitment* reflects whether abduction or other forcible recruitment strategies are ever employed during a given conflict. Data on rebel use of forced recruitment is taken from Cohen (2013a). One potential limitation of this indicator is that it only reflects forcible recruitment at the conflict level and does not explicitly identify whether a specific group relied on this strategy.[17] Regardless, to my knowledge it is the only readily available cross-national measure of such behavior covering the majority of groups included in this sample.[18]

Previous research suggests that some rebel groups may employ female suicide bombers because they provide the group certain tactical and strategic advantages. Particularly, female bombers are more likely to avoid scrutiny by security agents and may be more effective and more lethal in their attacks (Bloom 2011; O'Rourke 2009). For this reason, groups that routinely rely on suicide terrorism may also be more likely to recruit women and to deploy them in this role. It is worth nothing that not all groups that use suicide terrorism employ female bombers; moreover, some groups that utilize women in such roles also employ them in other types of combat roles. The variable *suicide terrorism* is a binary indicator reflecting whether or not the group used suicide bombings as a war strategy. Information for the measure is taken from the Chicago Project on Security and Terrorism (2015).

I include the variable *population size* because rebel groups active in countries with large populations should theoretically have a larger pool of recruits to draw from, thus potentially mitigating the need to recruit female combatants as conflict costs rise. In other words, a large pool of potential male recruits would reduce the pressures on reluctant rebel leadership to open combat positions to women. This variable is operationalized as a natural log of the average size of the population of the country in which the conflict took place during the years in which the group was active. The measure is adapted from data available in Gleditsch (2002).

The binary indicator *active 2000s* reflects cases ongoing after 1999. As shown in figure 4.2 above, the distribution of the prevalence of female combatants varies somewhat by decade. Most notably, the proportion of cases with high levels of female combatants marginally increases while the proportion with no evidence of participation appears to decline somewhat for conflicts surviving into the 2000s. I therefore include these indicators to help adjust for the differences across these time periods. Specifically, they account for potential bias caused by the possibility that more (or better) information is available for more recent conflicts. If this were the case, I would expect that groups that were active in more recent years would be more likely to be coded as having female combatants. However, it is also possible that global advancements in women's rights and status have also positively influenced the likelihood that rebels recruit women for combat roles. In either case, these variables should help capture this trend.

Finally, *fertility rate* is a measure of the number of live births per woman in a given country. I use this measure to capture the general status of women in society and the rights and privileges afforded to them. While this is a crude measure, it has been commonly employed in previous studies as a proxy for women's status in a society because it "captures multiple aspects of the complex matrix of discrimination and inequality" (Caprioli 2005, 169). High fertility rates indicate a social expectation that women's primary duties are childbearing and motherhood, and, in practice, inhibit women from participating in social and economic activities at rates comparable to men. Where fertility rates are high, women are less likely to be as educated as men, to work outside the home, and to engage in political activism. All else being equal, countries in which women face the greatest pressures to assume maternal roles at the expense of other positions in society are the least likely to produce armed groups that are willing to deploy women in combat. This indicator is taken from the World Bank's (2018) "World Development Indicators." Because the data on women's participation in armed groups are time invariant, I use the value at the start of the conflict. A country's fertility rate may vary as a function of the severity of the war, which is included in the model as an independent predictor. As the costs of war increase, the annual number of births is likely to decline due to the deterioration of the healthcare system, the displacement of the population, and the dissolution of family structures.[19] Using the value at the start of the conflict reduces potential endogeneity among these variables.

Analysis and Results

The dependent variables in the following analyses are the various measures of the presence and prevalence of female combatants contained in WARD. I rely on probit models to assess the probability of observing the presence or absence of female combatants in a rebel group (*female combatants*). In order to assess the prevalence of female combatants (*female combatant prevalence*) in a rebel group, I employ ordered probit models, which are a generalization of the probit regression that explicitly models the likelihood of observing one of multiple possible discrete outcomes on an ordinal scale.[20] I cluster standard errors on the country in which the conflict occurred in all models in order to correct for correlation among conflicts occurring within the same country.

I present the results from the regression analyses in table 4.2. The various models utilize the versions of the female combatant indicators specified at the top of each column. Models 1 and 2 report the result of the probit model using the binary *female combatants* indicator as the dependent variable. Models 3 through 7 include results from ordered probit models in which the various ordinal measures *female combatant prevalence* are employed as the dependent variable. Overall, the results provide support for the primary hypotheses related to the expected roles that conflict costs and rebel political ideology play in determining the prevalence of female combatants in an armed group. The variable *annual battle deaths*, which served as a proxy for group resource demands, is positive and statistically significant

Table 4.2 Factors Influencing Presence and Prevalence of Female Combatants

	Model 1	**Model 2**	**Model 3**	**Model 4**	**Model 5**	**Model 6**	**Model 7**
	Female combatants	Female combatants	Female combatant prevalence (best)	Female combatant prevalence (best)	Female combatant prevalence (high)	Female combatant prevalence (excluding suicide)	Female combatant prevalence (best)
Leftist	1.176 (0.326)**	1.321 (0.318)**	1.290 (0.316)**	1.488 (0.315)**	1.351 (0.340)**	1.439 (0.309)**	1.536 (0.334)**
Secular nationalist	0.295 (0.296)	0.101 (0.262)	0.062 (0.298)	−0.038 (0.252)	0.190 (0.289)	−0.070 (0.253)	−0.042 (0.257)
Religious	−0.141 (0.383)	−0.760 (0.431)*	−0.317 (0.367)	−0.654 (0.376)*	−0.920 (0.462)*	−1.097 (0.503)*	0.691 (0.680)
Annual battle deaths†	0.123 (0.068)*	0.186 (0.057)**	0.118 (0.058)*	0.164 (0.049)**	0.173 (0.049)**	0.199 (0.055)**	0.230 (0.060)**
Duration†	0.227 (0.082)**	0.258 (0.097)**	0.194 (0.077)**	0.159 (0.088)*	0.271 (0.099)**	0.241 (0.098)**	0.172 (0.088)*
Forced recruitment		0.417 (0.236)*		0.356 (0.207)*	0.452 (0.203)*	0.385 (0.210)*	0.329 (0.203)*

Suicide terrorism		0.906		0.335	0.499	−0.631	0.308
		(0.418)*		(0.305)	(0.304)	(0.396)	(0.312)
Population size†		−0.079		−0.036	−0.046	−0.038	−0.030
		(0.070)		(0.070)	(0.063)	(0.071)	(0.068)
Fertility rate		−0.115		−0.081	−0.129	−0.093	−0.079
		(0.068)*		(0.053)	(0.053)**	(0.056)*	(0.051)
Active 2000s		0.381		0.416	0.352	0.342	0.433
		(0.266)		(0.209)*	(0.206)*	(0.221)	(0.209)*
Religious*							−0.191
Annual battle deaths†							(0.090)*
Wald X^2	33.64	80.31	57.63	90.67	119.41	102.96	102.46
BIC	321.88	278.23	506.89	446.65	491.59	403.55	449.43
N (groups)	249	220	249	220	220	220	220

**$p < 0.01$

*$p < 0.05$ (one-tailed test)

†natural log

Note: Coefficients from probit (Models 1 and 2) and ordered probit (Models 3–7) models with standard errors (clustered on conflict country) in parentheses.

across all of the models. These results indicate that as the severity of the conflict increases, the prevalence of female combatants is expected to increase. This corresponds to one of the core arguments made in chapter 1: rebel leaders' decisions to deploy women in combat often come as a result of growing resource demands produced by the expansion of intensification of the conflict. This result is consistent with Hypothesis 1.

The results of the variables accounting for group ideology are also generally consistent with the hypotheses presented above. Because the ideology variables are mutually exclusive, the coefficients compare the effect of a given ideology to the excluded base category, which in all cases is *secular (other)*. Across the various models, *leftist* is statistically significant and positively signed. Moreover, the absolute value of the coefficient on this variable is much larger than that of any of the other binary variables, suggesting that it exerts the largest substantive impact of any of the ideology variables. I address the substantive effects of the variables in more detail below. For now, it is sufficient to say that the results suggest that groups embracing Marxist, Maoist, or other similar socialist ideologies are much more likely to recruit a larger proportion of female combatants than groups espousing other political ideologies. This result is consistent with Hypothesis 2.

By contrast, the signs of the coefficients for *religious* are negative across all of the specifications, and it is statistically significant in the majority of them. It is important to note that the two cases in which the variable fails to achieve statistical significance are those that do not include the control of suicide terrorism. In other words, after accounting for this variable, and acknowledging that this represents the only combat role available to women in many radical religious movements, the religious ideology variable becomes a significant predictor of both the absence of female combatants (Model 2) and a lower prevalence of female combatants overall (Model 4). This relationship is also apparent in the increase in the absolute value of the coefficient in Model 5, which uses the version of *female combatant prevalence* that excludes cases where women only served as suicide bombers. These results support the assertion that groups organized principally around religious beliefs and goals are comparatively less likely to recruit substantial numbers of female combatants. They therefore provide broad support for Hypothesis 3.

While the *religious* and *leftist* variables appear to exert a significant influence on the likelihood of observing female combatants, the variable *secular nationalist*, which reflects groups that were primarily organized around the interests of a specific ethno-national community but adopted neither a religious ideology nor a Marxist-oriented ideology, fails to achieve statistical significance in any model. Moreover, the sign on the coefficient reverses across the specifications. This result therefore suggests that nationalist rebel movements are no more or less likely than the *secular non-Marxist* rebel groups (the excluded category) to recruit female fighters. Moreover, the results also imply that they are comparatively less likely to recruit female fighters than leftist rebel groups but more likely to recruit female fighters than religious movements.

The absence of a statistically significant relationship between nationalist ideologies and female combatants may appear somewhat surprising. As I noted in chapter 1, some previous studies have asserted that nationalism inherently promotes patriarchy and forces women (and men) into traditional roles. Yet, as highlighted therein, nationalism often interacts with other political ideologies, producing a wide range of attitudes toward women's roles in combat. As such, nationalism itself may exert little independent influence on group attitudes toward the deployment of women in combat, and nationalist ideologies may be highly malleable with respect to attitudes toward gender norms and women's roles. As such, the prevalence of female combatants in these groups is likely to be shaped by the other ideologies that are often grafted onto nationalist movements. For instance, nationalist movements that also adopt Marxist ideologies (e.g., the EPLF, PKK, and ETA) are more likely to include female combatants, while those that incorporate orthodox religious beliefs (e.g., ASG or Hamas) are unlikely to deploy women in combat.

In order to more directly assess the role of nationalist ideologies on female combatant prevalence, I conducted a series of additional analyses. First, I replicated the models presented in table 4.2 but included only a variable for *nationalist* ideology in place of the other ideology indicators. This allowed me to determine if *nationalist* exerts any independent influence on the presence or prevalence of female combatants. Across each model in these alternative specifications, the variable is negative, but it does not achieve statistical significance. Second, I replicated the models, including the variables *leftist*, *religious*, and *nationalist* but interacted the nationalism indictor with each of the other ideology variables.[21] These interactions allowed me to explicitly examine the potential conditional effect of nationalism on the prevalence of female combatants in groups espousing either leftist or religious ideologies.

Plotting the marginal effects of the interaction demonstrates that nationalist ideologies do not moderate the relationships between the other ideologies and the prevalence of female combatants; rather, the relationship runs in the other direction.[22] However, the incorporation of leftist or religious ideologies substantially alters the likelihood that a nationalist group deploys female fighters. Specifically, nationalist groups that also adopt religious ideologies are less likely to utilize female fighters (e.g., MILF), while those that adopt leftist ideologies are more likely to employ women in combat (e.g., LTTE). This finding supports the assertion made in chapter 1 that the relationship between nationalism and women's roles in rebel movements is primarily driven by other ideologies that may interact with nationalist beliefs. It also provides further evidence that nationalist ideologies fail to exert an independent influence on the prevalence of female combatants.

Before proceeding to the discussion of Hypothesis 4, which posited that ideology conditions the relationships between conflict severity and female fighter recruitment, it is helpful to illustrate the substantial independent influence that ideology exerts on the prevalence of female combatants. Figure 4.5 depicts the substantive effects of the various ideology indicators on the probability of observing female combatants. Each panel in the figure illustrates the predicted probability of a given outcome of the four-category *female combatant prevalence* measure (*y*-axis) by the specified political ideology espoused by a group (*x*-axis), based on the results from Model 4. The most notable aspect of the figure is the large substantive effect exerted by the presence of a leftist ideology. With the exception of the top-right panel, which plots the predicted probability of observing a low prevalence of female combatants in a group, the effect of leftist ideology dwarfs all other ideologies. For example, in the top-left panel, the prediction suggests that groups espousing leftist political ideologies have just over a 10 percent probability of having no female combatants. By contrast, groups espousing a predominantly secular nationalist or other secular ideology have just over a 60 percent probability of observing this level of female fighters, while the probability of no female fighters in groups with religious ideologies is over 80 percent. The effect is equally striking in the bottom-right panel, which shows the probability of observing a high prevalence of female combatants. Here, the probability of observing that prevalence of female fighters is approximately 35 percent for leftist groups but only about 3 percent for both secular nationalist and other

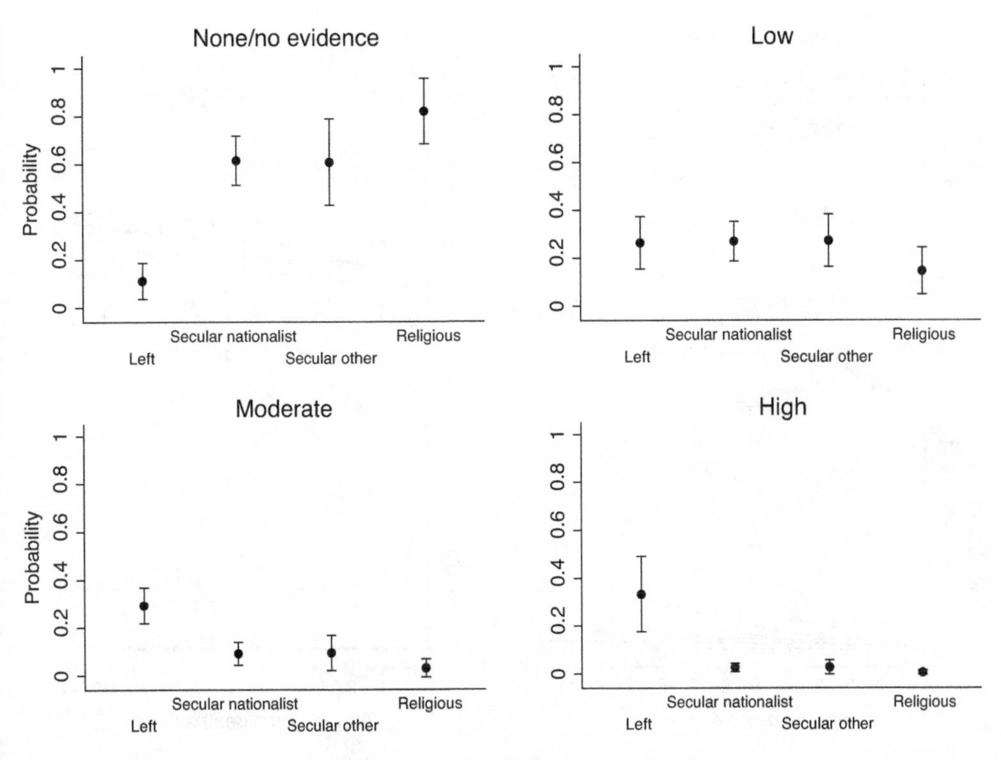

Figure 4.5 Predicted effect of group political ideology on female combatant prevalence

secular movements, and less than 1 percent for groups with religious ideologies.

Finally, Hypothesis 4 proposed that the type of ideology the group embraces conditions the influence of conflict severity on leaders' decisions to deploy women in combat roles. Rising conflict costs are expected to have a comparatively weaker influence on the decisions of the leaders of rebel groups adopting orthodox or fundamentalist religious ideologies than on groups espousing other ideologies. Thus, while secular rebel movements become increasingly likely to recruit and deploy female fighters as the resource costs produced by the conflict increase, those groups embracing religious ideologies are less likely to do so. In order to test this hypothesis, Model 7 includes a term reflecting the interaction of *religious* and *annual battle deaths*. In order to assess the significance of this proposed relationship and its substantive effect I present the predicted probabilities from that model in figure 4.6.

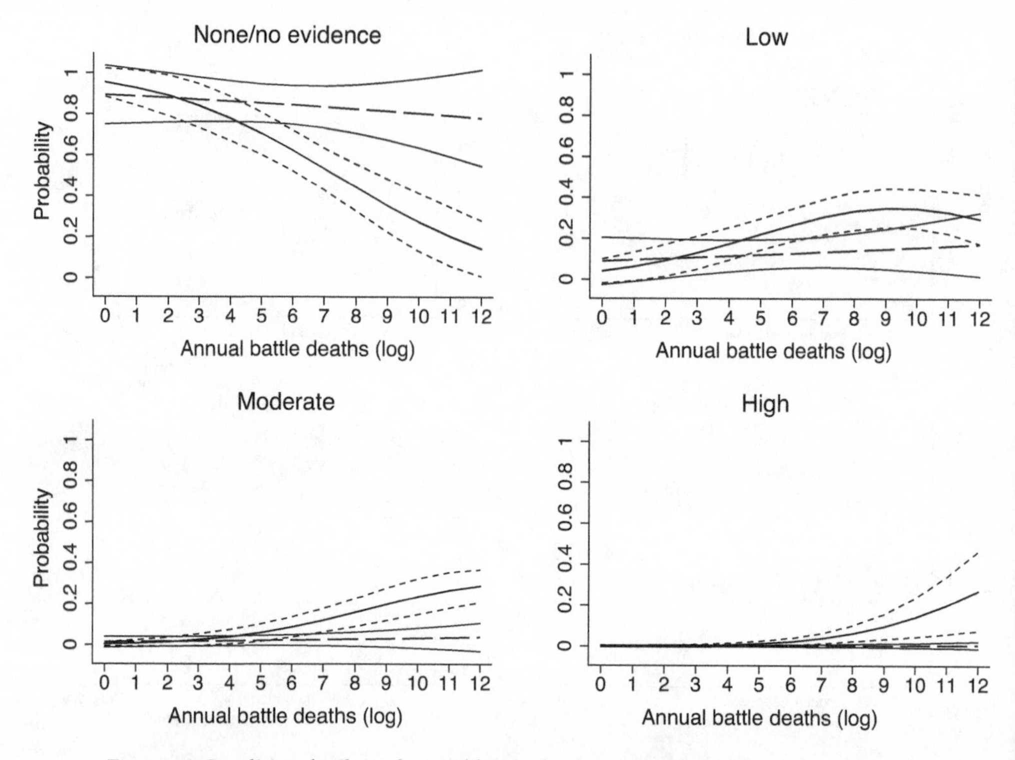

Figure 4.6 Conditional effect of annual battle deaths on female combatant prevalence

Each panel of the figure represents the probability of observing a given outcome of the four-category *female combatant prevalence* measure (*y*-axis) over the range of value of *annual battle deaths* (*x*-axis). The thick solid black and dashed black lines represent the predictions for groups espousing nonreligious and religious ideologies respectively. Examining the top-left panel of the figure, rebel groups that espouse religious ideologies as well as those that adopt nonreligious political ideologies both become less likely to completely exclude female fighters as conflict severity increases. However, the rate of the decline in these probabilities differs markedly over the range of the severity variable. For very minor conflicts, the difference between the two groups is minimal, and there is no statistically significant difference between them, as indicated by the overlapping confidence intervals. However, the difference between the two groups increases as conflict severity increases. At the mean level of annual deaths for conflicts in the sample (~6.5 on the logged scale), groups with religious ideologies are predicted to have an almost 85 percent chance of having no female fighters, while those with nonreligious ideologies have only a 60 percent chance of excluding female fighters. At one standard deviation above the mean (~8.5 on the logged scale), the likelihood of no female fighters declines to 35 percent for nonreligious groups but only to 80 percent for those with religious ideologies. Thus, while both groups become increasingly likely to recruit some number of female fighters as the conflict intensifies, the cost threshold at which rebel groups with nonreligious ideologies become willing to utilize female combatants is much lower than the threshold for religious groups.

The probabilities of observing a moderate prevalence of female combatants, which is depicted in the bottom-left panel, also illustrate this trend. For instance, religious and nonreligious groups are comparatively equally resistant to deploying substantial numbers of female combatants in minor conflicts. However, nonreligious groups become increasingly likely to do so as the conflict intensifies, while there is virtually no change in the odds that a religious movement utilizes female combatants. For a hypothetical conflict that is one standard deviation more severe than the sample mean (~8 on the logged scale), the probability of a nonreligious group employing a moderate prevalence of female combatants is approximately 15 percent, while the probability for a corresponding religious group is less than 4 percent. Moreover, for nonreligious groups this probability abruptly increases beyond this point, rising to almost 30 percent for the most intense conflicts. However, the probability of observing a high prevalence of female combatants remains essentially flat for armed groups embracing religious ideologies. These predictions suggest that group ideology—particularly the presence of a religious ideology—moderates the influence of conflict severity on the prevalence of female combatants. These predictions support Hypothesis 4.

Before concluding, it is worth reviewing the results of the control variables. First, consistent with the anecdotal evidence presented in chapter 1, the likelihood of observing female combatants is greater in longer conflicts compared to shorter conflicts. The variable *duration*, which reflects the logged value of the number of years the group was actively involved in armed conflict against the state, is positive and statistically significant across the specifications. Because the data employed in the analysis are time invariant, it is not possible to determine if rebel groups on average deploy women at a later date in the conflict. Nonetheless, these results indicate that the probability of observing female combatants increases as the duration of a conflict increases. They also indicate that factors such as group ideology and the intensity of the conflict are more influential in determining the prevalence of female fighters than the age of the group. For example, an increase in the logged duration of the conflict from the mean to one standard deviation above the mean decreases the probability of observing no female combatants from 88 percent to 84 percent and increases the probability of observing a high prevalence of female combatants from 2 percent to 3 percent. These changes are substantively much smaller than those resulting from comparable changes in the severity of the conflict or the differences between ideological categories.

It is also worth noting that group ideology conditions the relationship between conflict *duration* and *female combatant prevalence.* Much like the relation between conflict costs and the prevalence of female combatants, the likelihood that a group embracing a nonreligious political ideology utilizes female fighters increases as the duration of the conflict increases. By contrast, the likelihood that a religious rebel movement employs female fighters remains relatively flat regardless of the conflict's duration. This is largely consistent with the arguments put forth in chapter 1, though these results may reflect evidence of an alternative, complementary mechanism as play. Specifically, they suggest that while leaders of secular armed movements might initially resist incorporating women into the group's combat force, they become increasingly likely to do so as the conflict drags on and the resources necessary to sustain the movement continue to increase. However, consistent with the expectations of the theoretical arguments outlined in chapter 1, leaders of religiously motivated rebel movements continue to exclude women from combat roles.

The coefficient on the variable *forced recruitment* is positively signed and statistically significant in each of the specifications. These results support the claim that groups that rely on the forced recruitment of combatants are also more likely to deploy female fighters. This result is intuitive because groups that resort to involuntary recruitment methods to meet their resource demands are likely to be less selective in their recruitment. Indeed, such groups often utilize child soldiers, who are generally viewed as less effective combatants.[23] This result is also generally consistent with the logic that underlies one of the principal arguments outlined in chapter 1. Given that the need for human resource inputs is a primary driver of forced recruitment, it makes sense that groups that engage in this practice are also more likely to recruit female combatants. In other words, similar processes may drive both strategies.

Somewhat surprisingly, while *suicide terrorism* is positive in most models, it achieves statistical significance in only a single model. This result contrasts somewhat with previous findings, which often identified reliance on terrorist tactics as a primary motivator for recruiting women. It is perhaps worth reiterating here that while some groups employ women as suicide bombers and in other combat roles (e.g., LTTE and PKK), the majority of groups that employ them as suicide bombers restrict their participation to only that role (e.g., Hamas and Boko Haram). Indeed, in this sample only six groups employed women as suicide bombers but also allowed them to participate in other combat roles, while eleven employed them exclusively to conduct suicide bombings. Moreover, as Wood and Thomas (2017) note, this appears to be the only combat role that women are afforded in fundamentalist Islamist organizations, which are also the most likely to utilize suicide terrorism.[24] Thus, the effect of this variable on women's participation seems to be largely driven by the presence of a (very) small number of female suicide bombers in extremist Islamist rebel groups and not a general relationship between suicide bombings and female combatants.

Population size is negative across the models in which it is included, but it is not statistically significant in any of them. Thus, groups with access to larger populations are on average no less likely than other groups to recruit female combatants. Consistent with previous studies (e.g., Thomas and Wood 2018), *fertility rate* is negative and statistically significant in most of the specifications in which it is included. According to these results, as the number of births per woman in a given country increases, the likelihood of observing female combatants declines. To the extent that this measure captures information about the social expectations imposed on women and the gender norms of the society, this result suggests that the extant social status of women is likely to influence rebel leaders' willingness to deploy women in combat roles.

Lastly, the variable *active 2000s*, which indicates that a rebellion was active until at least the year 2000, is positive across the specifications and is significant in three models. Thus, those rebellions that were ongoing or began in the 2000s include greater numbers of female fighters than those that terminated in earlier time periods. This finding suggests that female fighters may have become more common in more recent years. Alternatively, it could suggest changes in news coverage, reporting, and general interest that have eased the task of identifying the presence of female combatants. In either case, the result indicates that the time period in which the conflict occurs is an important predictor of female fighter recruitment.

Conclusion

In this chapter I proposed a set of testable hypotheses based on the arguments set forth in chapter 1. I then outlined the data and methods used to assess the validity of those hypotheses and discussed the results of the analyses. Overall, the results reported above support the arguments made in chapter 1. As I argued, the decision to deploy women in combat roles (and the numbers in which to recruit them) rests with the groups' leadership. As such, the critical question is what factors are likely to positively incline the leaders of rebel movements to open recruitment to women and to train, arm, and deploy them in combat?

The argument I present suggested that a combination of strategic and ideological factors shape leaderships' decisions in this regard. More specifically, rising resource demands brought about by the intensification of conflict should, on average, increase leaders' willingness to accept women into the groups' armed wings as a way to address rising resource constraints. As the severity of the conflict increases, the prevalence of female fighters is therefore expected to increase. I further argued that the political ideology adopted by a group strongly influences the prevalence of female combatants. Revisionist and revolutionary ideologies are more likely to favor disrupting, challenging, and replacing extant social norms and orders, including those related to gender roles. By contrast, more conservative and traditionalist ideologies are more likely to favor preserving or reinforcing such norms and the structures that support them. An implication of this general relationship is that rebel groups that embrace leftist ideologies are much more likely to recruit and deploy female fighters, while fundamentalist and orthodox religious movements are much less likely to do so.

I empirically evaluated four hypotheses drawn from these arguments. Overall, the results presented above support these hypotheses. First, I find support for the general argument that sharply rising resource demands increase the probability of observing female combatants. Second, I find evidence that groups espousing leftist ideologies and those embracing religious ideologies represent opposite ends of the spectrum in terms of the prevalence of female fighters in their ranks. The former are much more likely to permit women to participate in combat, while the latter are much more likely to exclude them. I also find that groups with other ideologies, including nationalism and other forms of secular ideologies, occupy the middle ground between these poles. Furthermore, the results suggest that nationalism has no independent influence on the prevalence of female combatants. Nationalist groups are therefore no more or less likely than other groups to utilize female combatants. Rather, nationalism tends to intersect with other ideologies that do exert a significant influence on the prevalence of female combatants. Lastly, I found that group ideology—particularly the presence of a religious ideology, conditions the relationship between conflict severity and female combatants. In short, while nonreligious movements become more likely to mobilize women for war as resource constraints increase, armed groups espousing orthodox religious beliefs appear unwilling to adopt this strategy, even at the risk of jeopardizing the success of the rebellion.

CHAPTER 5

Empirical Evaluation of the Effects of Female Combatants

The results of the analyses presented in chapter 4 support the strategic logic of female recruitment I advanced in chapter 1, which linked war costs and the acute resource constraints they impose to rebel leaders' willingness to utilize female combatants. However, these analyses provide little insight into the impact that the incorporation of female combatants has on the groups that ultimately recruit them. Chapter 2 directly engaged the question of the strategic implications of female combatants, arguing that the decision to mobilize women for war, particularly in large numbers, can effectively increase the size of the rebel group's combat force. I further highlighted the potential role that the presence of female combatants can play in shaping observer attitudes toward the movements that employ them, and in assisting these groups in securing support from diaspora communities and transnational activist networks. In this chapter, I therefore evaluate the empirical validity of these arguments. In order to assess the hypotheses presented in chapter 2, I combine group-level analyses using information on female combatants taken from WARD, which I discussed in detail in chapter 4, with a novel survey experiment. In the following sections I describe the research designs used to validate these hypotheses and present the results. Overall, the findings support the claims advanced in chapter 2.

Research Design

Each hypotheses stated in chapter 2 requires a distinct analysis. In order to examine Hypotheses 5 and 7, which focused on the effects of female combatants on rebel troop mobilization and external support, respectively, I rely on conventional regression analyses of cross-sectional data that take the rebel group as the unit of analysis. In both cases, I use data from WARD to measure variation in the presence and prevalence of female combatants across the rebel groups in the sample. While WARD measures served as the dependent variables in chapter 4, in this chapter they become the primary independent variables of interest in models predicting the number of troops under a rebel group's command and the likelihood that a group successfully obtains support from transnational actors. I examine the validity of the relationship between external audience attitudes toward a rebel group and the presence of female fighters, which was formalized in Hypothesis 6, using data from a novel survey. Because each analysis employs different variables and utilizes different statistical methods, I discuss each in more detail below.

Recruitment, Replacement, and Rebel Military Capabilities

Hypothesis 5 asserts that rebel groups that incorporate larger numbers of female fighters into their ranks will be able to field larger and more capable combat forces compared to those that eschew women's participation. The logic for this argument is straightforward: increasing the pool of potential recruits should expand the number of troops under a group's command. To test this hypothesis, I rely on ordinary least squares (OLS) regression to evaluate the relationship between an estimate of rebel group troop size and a number of covariates. In this analysis, *rebel troop size* reflects the natural log of the estimated average number of troops available to a given rebel group as reported in the Non-state Actor (NSA) Dataset (Cunningham, Gleditsch, and Salehyan 2009, 2013). Because the hypothesis explicitly posits that increasing the proportion of female fighters in a rebel group should increase its overall size and material capability, I use the variable *female combatant prevalence* as the primary variable of interest in each model. As discussed in chapter 4, this categorical indicator represents a coarse estimate of the proportion of the rebel combat force that was composed of women.

In chapter 4 I also discussed the various alternatives measures of *female combatant prevalence* used in WARD, including both a "high" estimate and a measure excluding cases in which female fighters were strictly limited to suicide bombings (e.g., Hamas, the PIJ, and Boko Haram). In addition to the "best" estimate of this indicator, I also assessed models using these alternative estimates. Assessing the influence of the measure that excludes cases where female combatants were restricted to only the role of suicide bombers is important because the inclusion of these cases might bias the results against the hypothesized relationship. Specifically, suicide bombers are comparatively rare, representing only a handful of any given group's overall troop numbers. The inclusion of women in only these roles would therefore result in a negligible expansion of the overall force, though it might spur recruitment overall by shaming reluctant men to join the movement.

I also account for a number of confounding variables in the models. First, I account for whether or not the group was involved in a *secessionist conflict* with the state. Such conflicts typically pit an aggrieved minority ethnonationalist group against the state. Previous studies suggest that the combination of shared cultural identity and collective grievances facilitates political mobilization and that political conflicts centered on identity issues escalate more rapidly (e.g., Eck 2009; Gurr 1993). As such, rebel groups involved in secessionist conflicts have a built-in mobilization advantage, which may allow them to field more troops than rebellions that mobilize along other dimensions. On the other hand, because the autonomy-seeking group is often a numerical minority in the state, they might ultimately field smaller forces even if they are more easily able to mobilize troops. I am thus agnostic about the effect of this variable. It is a binary indicator coded 1 if the conflict focused on claims of territorial autonomy, and coded 0 if otherwise. Data for this indicator come from the UCDP Dyadic Dataset (Harbom, Melander, and Wallensteen 2007).

I also include the variables *population size* and *GDPpc* to account for state-level structural factors that might influence rebel mobilization capacity. I include the former because rebel groups involved in conflicts in countries with large populations should theoretically have access to larger a pool of potential recruits. As such, these countries might be more likely to host larger rebel movements. The latter variable is included because wealthier states possess greater resources with which to suppress rebellion and/or to address the grievances that drive it. Both mechanisms should theoretically suppress the supply of available recruits. Wealthier countries should therefore on average host numerically smaller and comparatively less capable rebel groups. The respective variables reflect the natural log of the average size of the population and average estimated per capita GDP of a given country during the years in which the group was active. Data for both variables are taken from Gleditsch (2002).

I control for the state's political institutions because repressive, autocratic states are more likely to create grievances among the population and to deny them legitimate means through which to address those grievances compared to democratic states. The presence of democratic institutions should help minimize support for the rebellion because individuals can express their grievances at the ballot box rather than in the risky act of rebellion. Much as rebellion is less common in democracies, rebellions that occur in democratic states should be comparatively weaker than those that arise in autocratic states. The binary variable *democracy* indicates whether or not the country in question had a (nominally) democratic political structure for the majority of the years during which the group was actively involved in a violent conflict with the state. I use the *democracy* variable from the "Democracy and Dictatorship" dataset (Cheibub, Gandhi, and Vreeland 2009). According to this measure, democratic states hold routine and competitive direct or indirect multiparty elections for the executive and legislature. I create this variable by taking the modal value of the government's democracy score for the years in which the group was involved in armed rebellion against the state.[1] I also include a control for the size of the state military force. *State troop size* reflects the natural log of the average estimated number of personnel in the state's military force during the years in which it was involved in armed conflict with the rebel force. This variable is taken from the National Material Capabilities Dataset, v. 5.0 (Singer, Bremer, and Stuckey 1972).

Forced recruitment is a binary indicator accounting for whether or not abduction or other forcible recruitment strategies were ever employed during the conflict in which the group fought. I include this measure because the use of these tactics is explicitly intended to increase the number of troops available to rebels and is frequently employed when the resource demands experienced by the group outpace the inflow of voluntary recruits. As such, it represents an alternative (or perhaps complementary) strategy to female recruitment. Moreover, as noted in chapter 4, groups that engage in forced recruitment are more likely to include female combatants. Including this variable helps to account for the possibility that forced recruitment rather than the decision to recruit female combatants drives any observed relationship between the prevalence of female combatants and the military strength of the rebels. Overall, and to the extent it is successful, rebels that engage in forcible recruitment should be able to field larger numbers of troops than those that do not. As discussed in chapter 4, this measure is adapted from Cohen (2013a).

The binary variable *rebel external support* indicates whether or not the rebel group under observation received military or nonmilitary support from a foreign state at any point during the conflict. These types of support include the provision of troops, military hardware or weapons, financial support, logistical support, the use of territory, and other types of assistance. I include this variable because external support can provide rebels with the equipment, knowledge, and space to recruit, train, and arm troops. These assets should facilitate the expansion of the rebel force. I likewise include the measure *transnational constituency*, which indicates that a nonstate transnational entity provided some form of tacit or explicit support for the rebel group.[2] These constituencies are often diaspora communities that fund-raise on behalf of their ethnic kin and provide them additional tangible and intangible benefits. However, they also include global religious communities and international solidarity networks (Salehyan, Gleditsch, and Cunningham 2011).[3] Groups that received support from such actors are coded as 1, and groups that received no such support are coded as 0. Both indicators of external support are taken from the NSA Dataset (Cunningham, Gleditsch, and Salehyan 2009, 2013)

Finally, I include a control for the length of time the group was actively involved in violent conflict with the state. The variable *duration* is the log-transformed value of the number of years between the group's initiation of violent conflict with a government and its cessation of those hostilities as reported in the UCDP Dyadic Dataset. I include this variable because rebel groups tend to begin life from a position of relative weakness and expand their capabilities over time (provided they survive) (e.g., Bapat 2005; Cunningham, Gleditsch, and Salehyan 2009). Thus, duration should be positively correlated with a group's overall size and mobilization capacity.

I present the results from a series of OLS models in table 5.1. Overall, these results support the contention that rebel groups that recruit female combatants ultimately field larger combat forces than those that exclude them. Across the models, the coefficients on the various versions of *female combatant prevalence* are positive and statistically significant, suggesting that groups including larger proportions of female combatants on average have larger numbers of troops than those that have fewer or no female combatants. This result is intuitive since removing artificial constraints on the supply of troops should facilitate mobilization. Assuming that significant numbers of male recruits do not eschew recruitment explicitly because the group contains female combatants, a group composed of 10 percent female combatants would arguable have foregone (more or less) that proportion of their force had they decided against recruiting women into the group's combat force.

It is possible that some number of male recruits might abandon the group or refrain from joining due to the presence of female fighters in the ranks. Yet, the anecdotal evidence discussed in chapters 2 and 3 suggests that, if anything, the presence of female combatants serves as an important recruitment device for armed groups. There is little evidence for a contrary, dampening effect of female combatants on recruitment. In addition to their direct influence on rebel troop size, female combatants are also likely to indirectly benefit rebel recruitment efforts. Principally, the visible presence of some number of women in combat roles—even in relatively small numbers—is likely to compel reluctant men to join the rebellion. Their presence may also exert a demonstration effect that generates additional female recruitment. Regardless of the specific mechanism, these results suggest that the decision to recruit female combatants on average increases the number of troops available to the rebel movement. This provides support for Hypothesis 5.

In order to better understand the substantive impact of *female combatant prevalence* on *rebel troop size*, I present the predictions from Model 4 in figure 5.1.[4] The figure illustrates the predicted value of *rebel troop size* (y-axis) over the categories of *female combatant prevalence* (x-axis). The results suggest that a one-category increase in the measure of female combatant prevalence is expected to produce a change of 0.244 in the logged troop estimate. An average group with no female combatants would be expected to have an estimated logged troop value of approximately 8.04 (~3,100 troops); however, increasing female combatant prevalence to the low, moderate, and high categories increases the expected troop size to roughly 8.29 (~4,000 troops), 8.53 (~5,100), and 8.78 (~6,600), respectively. These results

Table 5.1 Effect of Female Combatants on Rebel Troop Size

	Model 1	Model 2	Model 3	Model 4
Female combatant prevalence (best)	0.370 (0.098)**	0.224 (0.091)**		
Female combatant prevalence (high)			0.252 (0.086)**	
Female combatant prevalence (excluding suicide bombers)				0.244 (0.091)**
Secessionist	0.495 (0.264)*	0.011 (0.294)	−0.045 (0.289)	0.012 (0.290)
Population size†	−0.301 (0.113)**	−0.198 (0.108)*	−0.186 (0.107)*	−0.192 (0.105)*
GDPpc†	−0.558 (0.141)**	−0.529 (0.155)**	−0.548 (0.151)**	−0.518 (0.151)**
Democracy	−0.629 (0.241)**	−0.550 (0.286)*	−0.532 (0.282)*	−0.561 (0.285)*
State troop size†	0.246 (0.108)*	0.128 (0.087)	0.127 (0.086)	0.127 (0.086)
Forced recruitment		0.421 (0.219)*	0.381 (0.213)*	0.426 (0.219)*
Rebel external support		0.619 (0.199)**	0.627 (0.195)**	0.608 (0.199)**
Transnational constituency		0.445 (0.222)*	0.462 (0.219)*	0.467 (0.221)*
Duration†		0.236 (0.093)**	0.212 (0.096)*	0.225 (0.092)**
F	6.09	12.26	12.49	12.12
R^2	0.22	0.37	0.38	0.37
BIC	747.97	655.07	650.90	653.76
N (groups)	210	192	192	192

** $p < 0.01$

* $p < 0.05$ (one-tailed test)

†natural log

Note: Coefficients from ordinary least squares with standard errors (clustered on conflict country) in parentheses.

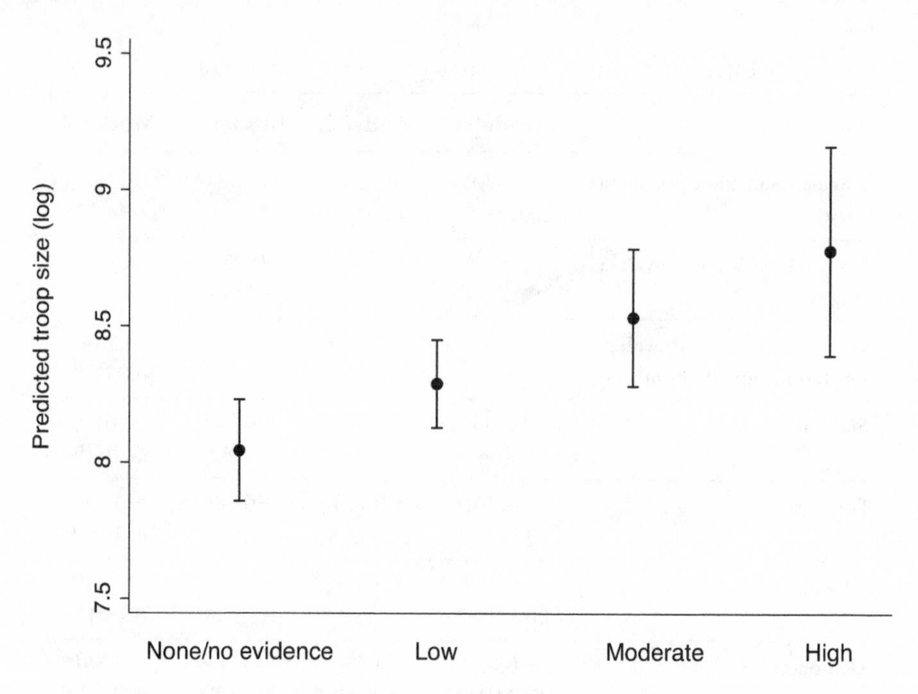

Figure 5.1 Predicted effect of female combatant prevalence on rebel troop size

imply that the effect of incorporating female combatants on rebel troop size is nontrivial. Consistent with the motives that prompted their recruitment, female combatants appear to produce more robust rebel groups, at least in terms of troop size.

While figure 5.1 shows the predicted changes in troop values over the categories of *female combatant prevalence*, it reveals little about the effect of recruiting female combatants on overall rebel troop strength relative to other recruitment strategies. As I argued in both chapters 1 and 2, recruiting female combatants represents one of several potential mobilization strategies by which rebel leaders seek to meet their human resource needs, expand the size of the rebellion, and increase the odds that the group achieves its political objectives. Forcible recruitment represents another strategy rebel leaders routinely employ to acquire human resource inputs during periods of excess demand.[5] As such, it is informative to compare the effect of forced recruitment and female combatant recruitment on rebel troop size.

In order to evaluate the comparative effects of these strategies, I rerun Model 4 but replace the prevalence variable with a binary variable indicating whether or not the group utilized female combatants. By using binary indicators as proxies for each recruit strategy, I can more easily compare their effects on troop size. The results imply that the effects are very similar: the coefficient for the binary *female combatants* variable is 0.490 (p = 0.002) while the coefficient for *forced recruitment* is 0.419 (p = 0.028). Furthermore, the predictions from the model suggest that groups that engage in forced recruitment have approximately 5 percent more troops than those that do not utilize this strategy.[6] By comparison, groups that recruit female combatants have approximately 6 percent more troops than those that fully exclude women from combat. Thus, the substantive effect of these two recruitment decisions is quite similar. Recall, however, that the results presented in the last chapter suggest that forcible recruitment is correlated with the prevalence of female combatants. This highlights the potential similarity in motivations behind the strategies. Rebel movements recruit female combatants and engage in forcible recruitment in order to secure human resources, and these results suggest that both represent nominally effective strategies in this respect. Nonetheless, the evidence suggests that they represent two distinct strategies, and the presence of female combatants is not contingent on the use of forcible recruitment.[7]

Turning to the other control variables, *secessionist conflict* is significant in a single model and switches signs across the specifications. This suggests that groups engaged in separatist conflicts are numerically no larger and no more militarily capable than their center-seeking counterparts. *GDPpc* is also negative across the specifications and statistically significant throughout. This implies that rebellions in wealthier states are on average smaller and less militarily capable than those in poor states. This result is not surprising given that wealthier states are more likely to possess the resources to effectively suppress rebellions and can devote substantial resources to addressing the grievances that drive political violence and incentivize rebel recruitment in the first place (should they choose). *Democracy* is also negative and significant across the models, suggesting that rebels challenging nominally democratic regimes are generally numerically smaller than those fighting nondemocratic governments. As with the GDP indicator, the result is intuitive.

By contrast, *population size* is surprisingly negative across the specifications. It achieves significance in each of the models. While the result is unexpected, it is possible it obtains because larger populations allow governments, which are often better able to mobilize resources (often by compelling service), to expand their military capacity rapidly when needed. If this were the case, we would expect rebel forces to be comparatively smaller in states with larger militaries. In other words, states could convert population size into coercive power, thereby suppressing rebellion. The results, however, do not appear to support this claim. *State troop size* is positive across the models but only achieves statistical significance in one of them, suggesting that the size of the state's military has limited influence on the size of the rebel group. This ambiguous relationship likely reflects the interdependent nature of the variables: states increase the size of their military in response to rising rebel threats.

The results suggest that external support provided either state or non-state actors is positively correlated with the size of the rebel group. Moreover, *rebel external support* and *transnational constituency* are both significant across the models in which they are included. These results are largely anticipated since groups that receive material support are more likely to possess resources that enable them to effectively recruit from the population. Previous studies have also proposed that a group's ability to secure support from external actors—particularly international IOs and NGOs—is often perceived as a signal of legitimacy by domestic constituencies (e.g., Bob 2005). To the extent this is true, groups with such support may enjoy superior mobilization capabilities, resulting in larger troop sizes. Lastly, *duration* is positive and significant across the models, suggesting that the size of rebel movements tends to increase over time—or at the least that longer-lived rebellions field larger numbers of troops. This is unsurprising given that potential supporters are more likely to mobilize in support of a group once it has demonstrated its ability to survive and impose costs on the state.

These results provide support for Hypothesis 5 and are consistent with the primary arguments made in chapter 2: rebel groups that recruit female combatants are better able to address their human resource needs and ultimately are able to amass larger combat forces than those that exclude women. An admitted limitation of this analysis is that the data are not time-series and therefore cannot accurately assess the causal relationship proposed in Hypothesis 5. Specifically, it is possible that the groups in the sample that ultimately elected to recruit female fighters already fielded larger than average combat forces. If this was the case, the relationship observed here might be spurious or stem from endogeneity among the variables. In light of this limitation, these results should be treated as preliminary and should be reevaluated in the advent of data that can better account for the causal ordering posited in the argument. Nonetheless, they are consistent with both the argument presented in chapter 2 and the anecdotal evidence presented in chapter 3.

In the following sections I address additional arguments made in chapter 2. Specifically, I evaluate the effect of female fighters on audience attitudes toward the groups that employ them. I also consider whether the presence of female fighters increases the likelihood that a group actually receives support from external actors.

Female Fighters and Audience Attitudes

Hypothesis 6 in chapter 2 asserted that the presence of female fighters increases domestic and international audiences' overall sympathy and support for the rebels and their goals. In order to assess the validity of this claim, I rely on evidence from a survey experiment intended to examine the effect that awareness of the presence of female combatants in a hypothetical armed group exerts on a respondent's perceptions of the group. The instrument used in the experiment was designed in collaboration with Devorah Manekin, to whom I am indebted for allowing me to share a portion of our work here. The survey experiment was deployed in multiple waves and to multiple samples of respondents.[8] A pilot version was deployed in spring 2016 to a sample of 450 undergraduate students enrolled in online classes at Arizona State University (ASU). After making very minor modifications in the wording of the treatment used in the pilot study, the survey was again deployed to a national sample of more than 1,800 online respondents in summer 2017.[9]

While neither sample is completely representative of the demographics of the U.S. population, both nonetheless capture substantial variation in the demographic indicators normally viewed as relevant to experimental studies of this type (e.g., gender, age, race, and so on).[10] Most importantly, in each wave of the survey the treatment was randomized, ensuring that the individual characteristics of the respondents had no influence on the likelihood that a given respondent received the treatment. Moreover, in both cases there is nominal balance across the relevant demographic categories, indicating that a roughly equal number of male/female, white/nonwhite, and liberal/conservative respondents received the treatment versus the control. Because the experiment utilizes a convenience sample of U.S. respondents, the results are not necessarily representative of populations in other locations. They nonetheless suggest that respondents viewed the hypothetical armed resistance movements somewhat more favorably when they were made aware that the group included a substantial number of female combatants. This result provides some preliminary support for arguments regarding female combatants and (external) audience support.

The experiment consisted of a simulated news article describing a fictional conflict involving an armed self-determination movement named the Kaftarian National Front (KNF). After completing a battery of demographic and individual difference measures and reading introductory text describing the survey and providing instructions, all participants received the same narrative describing a rebel group attempting to establish an independent homeland. Specifically, subjects received the following description of the conflict:

> For several months, the Kaftarian National Front (KNF), a separatist movement located in South Asia, has been engaged in a violent struggle against the government that has claimed hundreds of lives on both sides. The conflict has rapidly escalated in recent months, and experts believe it has the potential to spill over into neighboring countries, potentially destabilizing the region and creating severe humanitarian costs.
>
> Human rights organizations have expressed serious concerns over the deteriorating human rights situation in the country. They accuse the government of committing many serious human rights abuses and using harsh counterinsurgency tactics that violate international humanitarian law, including the execution of civilians suspected of supporting the KNF and the indiscriminate bombing of villages in KNF-held areas.

Participants were then randomly assigned to an article that described either a mixed-gender rebel group in which one-third of the fighters were reportedly female or to a nearly identical article including only references to male combatants. Nearly identical photographs of a group of fighters accompanied the articles. Specifically, the photograph accompanying the article describing the mixed-gender group included a female combatant in the frame, while the female fighter was cropped from the frame of the photograph included in the all-male article. Thus, the photographs provided to the treatment and the control groups differed only in terms of the inclusion of a female combatant.

Compared to the argument put forth in chapter 2, the treatment used in this experiment is extremely subtle. Indeed, with the exception of hair length and facial features, the male and female combatants in the photos are extremely similar. Both assume the same poses, have similarly stoic expressions on their faces, carry the same weapons, and wear the same uniforms. Yet, propaganda materials and media coverage are often intended to promote audience interest, and the former are explicitly intended to shape the attitudes of the observer. As discussed in chapter 2, such materials therefore highlight the exceptional nature of women's participation in combat and attempt to frame their participation in ways that maximize its impact on audience perceptions, including explicit efforts to accentuate and discuss the feminine characteristics of female fighters. The experiment used here provides the audience only with knowledge that female fighters participated in the rebel force in substantial numbers and provides an image including a single female combatant. It thus allows respondents to interpret their participation independently. The virtue of this more subtle treatment is that it permits a clear test of the underlying impact that the simple presence of female fighters has on audience attitudes rather than the effect of any additional persuasion or contextualization.

Hypothesis 6 posits a direct and unconditional relationship between female combatants and audience attitudes; however, it is possible that different characterizations of the movement (and the combatants included in it) also shape audience attitudes. Particularly, as discussed in chapter 2, the presence of female combatants might ameliorate some of the adverse effects of negative characterization of the group. This would be indicative of success in rebel efforts to create a more sympathetic counternarrative in the face of government efforts to characterize the group as nothing but thugs and terrorists. In order to examine this conditional effect, the survey experiment included an additional manipulation. In addition to random assignment to gender treatment condition, respondents were also randomly assigned to one of two conditions describing rebel interactions with the civilian population.[11] In the first, the rebel group was described as having positive relations with civilians, including building schools and operating medical clinics. In the second, the group was characterized as abusive toward civilians, engaging in looting, forced recruitment, and other abuses. The addition of this condition provides insight into whether negative descriptions of the group's behavior condition any effects that the presence of female combatants might exert on audience perceptions of it.

The dependent variable used in the analyses is a categorical measure of respondents' self-reported support for the group and its goals. *Support* is a five-point categorical measure based on responses to the following question: "Based on the information you read, to what extent do you support or oppose the KNF and its goals?" The scale provided to the respondents ranged from "strongly support" to "strongly oppose." The variable is reverse coded in the analysis for ease of interpretation: higher values of the dependent variable reflect greater support.

The variable reflecting the treatment condition for female combatants is labeled *mixed gender*, while the condition describing whether or not the group abused civilians is labeled *abusive*. Because I am primarily interested in the effect of the gender treatment, I focus primarily on it and view the other treatment as a moderating variable that potentially conditions the relationship between the gender treatment and audience attitudes. I first present a simple difference of means tests for *mixed gender* and then present results from a series OLS models.[12] I include both treatment conditions in each of the regression models. In addition, in one specification for each sample I include an interaction term created by multiplying the two treatment variables. In these models, I also include controls for relevant demographic and attitudinal characteristics that might influence respondent support for the group. The first of these is respondent *sex*, which is coded 1 if female, and 0 otherwise. I include the variable *race*, which is coded as 1 if the respondent identified as white and 0 if they identified as any other racial group. *Age* is a categorical indicator coded by decade. *Education* is a six-category indicator ranging from "did not complete high school" to "post-graduate degree." Because the student sample is composed only of respondents enrolled in university courses, I exclude this measure from the analyses of the pilot study. *Liberal* is a binary indicator denoting that the respondent identified as liberal.[13] *Income* is a four-category indicator reflecting self-reported household income ranging from less than $25,000 to greater than $75,000.[14]

The results from the experiment, which I present in table 5.2,[15] provide preliminary support for one of the central arguments explicated in chapter 2: female fighters positively influence audience attitudes toward the

Table 5.2 Effect of Female Combatants on Audience Attitudes

	Student sample			National sample		
	Model 7	Model 8	Model 9	Model 10	Model 11	Model 12
Mixed gender	0.327 (0.102)**	0.317 (0.100)**	0.103 (0.140)	0.086 (0.052)*	0.086 (0.049)*	0.092 (0.070)
Abusive		−0.485 (0.100)**	−0.696 (0.140)**		−0.609 (0.049)**	−0.604 (0.070)**
Mixed gender* abusive			0.426 (0.198)*			−0.012 (0.099)
Sex (female)		0.088 (0.104)	0.080 (0.103)		−0.036 (0.051)	−0.036 (0.051)
Race (white)		−0.077 (0.104)	−0.071 (0.102)		0.079 (0.057)	0.079 (0.057)
Age		−0.011 (0.073)	−0.010 (0.073)		−0.056 (0.015)**	−0.055 (0.015)**
Education					0.022 (0.020)	−0.022 (0.020)
Liberal		0.197 (0.100)*	0.198 (0.100)*		0.104 (0.053)*	0.104 (0.053)*
Income		−0.005 (0.042)	−0.003 (0.041)		−0.019 (0.024)	−0.019 (0.029)
F		5.86	5.76		21.48	18.09
R^2		0.09	0.10		0.09	0.09
BIC		1,280.5	1,281.85		5,098.69	5,106.58
N	429	429	429	1,741	1,741	1,741

*$p < 0.05$ (one-tailed test)

Note: Difference in sample means (t-test) (Models 7 and 10) and coefficients from ordinary least squares regression models (Models 8–9 and 11–12) with standard errors in parentheses.

groups that employ them, making them more supportive of the group and its goals. Models 7 through 9 report results for the pilot study conducted on the student sample, while Models 10 through 12 report the results for the larger national sample. The results of both simple t-tests (Models 7 and 10) and the OLS models including a number of control variables (Models 8 and 11) reveal a positive and statistically significant treatment effect for *mixed gender* in both the student sample and the national sample. In both samples, the respondents who received the treatment condition highlighting the presence of female combatants in the hypothetical armed group were more favorably inclined to the group and its goals compared to those who received the description of the group that *did not* reference female combatants. Overall, the results therefore support the argument that the presence of female combatants can positively influence audience attitudes toward the group and its goals.

The other treatment condition, *abusive*, is also statistically significant across the models. This result (unsurprisingly) indicates that respondents were less likely to express support for the hypothetical armed group when it was described as acting violently toward the local population. Interestingly, the effect of the interaction of the treatment conditions varies across the two samples (Models 9 and 12). In the student sample, the interaction is significant and positive, while it is negative and insignificant in the national sample. These results therefore provide mixed evidence for the potential influence of female combatants to ameliorate the adverse influence of negative characterizations of the group on observer support for it.

In order to more clearly illustrate the effect of the treatments and to more closely examine the conditional effect of abuse on the relationship between the presence of female combatants and respondent attitudes, I present the marginal effects from the models in figures 5.2 and 5.3. Figure 5.2 shows a direct relationship between the presence of female combatants and respondent support for the group and its goals based on the results from Models 8 and 11. The figure illustrates the effect of *mixed gender* (y-axis) on the level of the respondent support (x-axis). The left panel of the figure shows the effect within the student sample while the right panel shows the effect within the national sample. As noted above, in both samples the treatment effect is significant, with those respondents exposed to the conditions highlighting the presence of female combatants expressing greater support for the group and its goals. However, the scale of the effect differs across the samples. In the student sample, exposure to the treatment increased support by roughly a third of a point (on the five-point scale) while in the national sample it increased respondent support by only approximately one tenth of a point. While the effect size is marginal, the effect is nonetheless discernible. Moreover, the effect size is both unsurprising and largely consistent with the argument.

In a real-world setting, respondents are exposed to the presence of women in war zones only in passing or only for brief moments, such as when they skim a news article online or watch a segment on cable news. A single, limited instance of exposure to an image of a female combatant would not be expected to produce a substantively large effect on respondent sentiment. However, repeated exposure to such images might have such an effect. Indeed, if the initial exposure to such images or information pique observer interest or provoke sympathy, it may encourage that observer to seek out additional information on the case or to devote greater attention to subsequent media reports about it. In this sense, initial exposure may lead to additional exposure, which may in turn lead to a deepening of the initial sympathetic sentiment. The design of this study prevents testing this speculative corollary hypothesis, but attention to this question

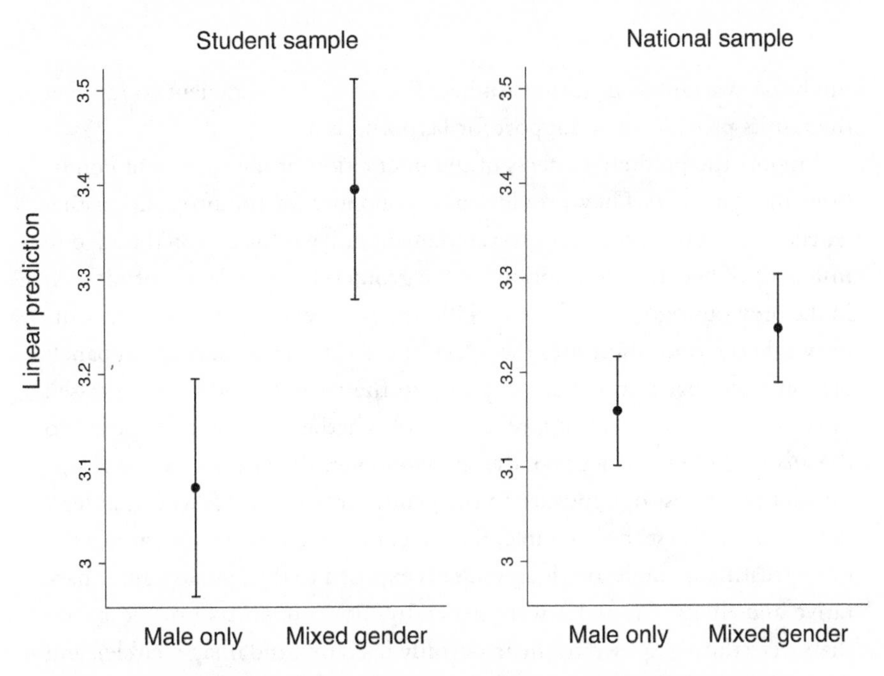

Figure 5.2 Predicted effect of mixed-gender treatment on respondent support

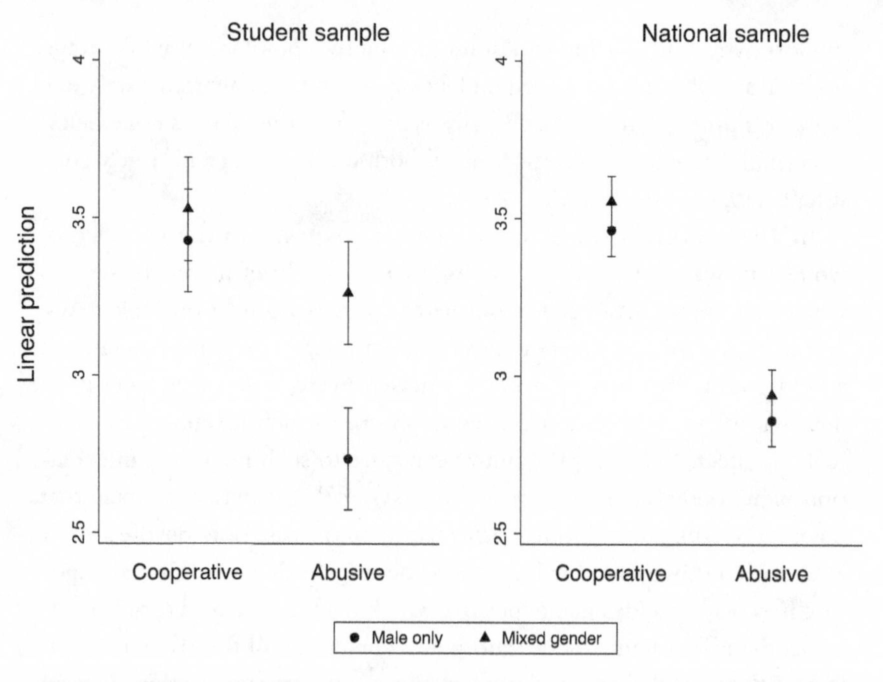

Figure 5.3 Conditional effect of mixed-gender treatment on respondent support

might be warranted in future studies. For now, it is sufficient to say that the results provide some support for Hypothesis 6.

I report the predicted effects of the interaction of the treatment conditions in figure 5.3. They provide mixed support for the proposition that female combatants exert an observable ameliorative influence on the adverse influence of negative descriptions of the group on respondent attitudes. As in the previous figure, the left panel illustrates the effect in the student sample while the right illustrates the effect in the national sample. Both panels demonstrate that respondents exposed to the *abusive* condition expressed lower support for the group, regardless of whether they were exposed to the *mixed gender* treatment condition. Moreover, the effect of abuse is significant regardless of exposure to the gender treatment. However, at least in the case of the student sample, the effect of *mixed gender* varies across the other treatment condition. Respondents exposed to the female fighter narrative and image (triangle) were generally more supportive of the group than were those exposed to the male-only narrative and image (circle), and the difference between the two is substantially larger in the presence of the abuse condition. Specifically, while the description of abuse drove down support, for both all-male and mixed-gender rebel groups, the decline is much steeper in the case of the hypothetical all-male group. In other words, there is some evidence that female fighters offset the negative characterization of the group. This effect is not apparent in the national sample. Consequently, these results provide only limited support for this assertion. Nonetheless, as a whole the results provide support for the arguments presented in chapter 2.

Securing External Support

Hypothesis 7 asserts that rebel groups that include female fighters are more likely to receive support from external actors than those that exclude women. The argument that produced this hypothesis focused primarily on the ability of female fighters to shape external audience perceptions of the group and its goals. This claim received some support from the previous set of analyses. Rebels often strategically highlight the presence of female fighters in their ranks in their propaganda materials and public relations efforts. More specifically, by focusing on (and sometimes exaggerating) women's participation in the group, rebels hope to present a more positive image to both domestic and international audiences and to counter the government's ubiquitous efforts to characterize its goals as illegitimate and to portray the movement as a band of thugs and terrorists. Additionally, owing to their (perceived) novelty, the presence of female fighters is more likely to attract attention from these audiences, thereby increasing the visibility of the group and its goals. To the extent that the presence of female fighters assists rebel groups in garnering sympathy, interest, and support for its goals from outside observers, it might also increase the odds that the group receives tangible benefits or tacit support from transnational actors such as diaspora communities, NGOs, or other solidarity organizations.

To my knowledge there has been no effort to quantitatively assess the factors that predict transnational nonstate actor support for rebel groups.[16] Moreover, few cross-national measures of external nonstate actor support for rebel movements exist. However, the NSA Dataset includes the variable *transnational constituency*, which I discussed briefly above. Given the dearth of systematic data on this topic, this variable arguably represents the best available proxy for the phenomenon of interest. In the following analyses, I use two versions of the basic variable. The first accounts for both tacit and explicit support. Tacit support generally includes statements of sympathy or solidarity by a community or activist group and may include lobbying, protests, and other demonstrations of support on behalf of the group. Explicit support indicates that transnational actors provided material support to the rebels, such as economic funding, supplies (e.g., food, medical supplies, clothing, and weapons), or fighters. The variable *transnational constituency* represents both tacit and explicit support and is coded 1 if the group under observation received any support from a diaspora community, transnational advocacy network, or other similar organization during its campaign against the state, and 0 otherwise. The related variable *transnational constituency (explicit)* is coded 1 for those cases that received explicit support, and 0 otherwise.

I also estimate one model that explicitly examines the effect of female combatants on the provision of material support to the rebellion by diaspora communities. While few good indicators of such support exist (presumably because such support is often covert), the UCDP External Support Dataset, v. 1.0 (Högbladh, Pettersson, and Themnér 2011) provides some information about such support. Using information from the dataset, I construct the variable *diaspora support*, which is coded as 1 for any group that received tangible support from a diaspora community composed of the same ethnic or national group as the rebels, and 0 otherwise. Because diaspora groups are almost exclusively composed of defined ethnic, religious, or national communities, I limit the sample of cases in this model to rebel groups advancing nationalist claims.[17] This eliminates the large number of cases for which I would not anticipate the presence of a diaspora community of any kind.

The primary independent variable in these analyses is the binary *female combatants* measure. I present results using two versions of the measure, one that includes cases of female suicide bombers and one that excludes them. I include both because much of the previous research on this topic has focused heavily on the media coverage of female suicide bombers and audience responses to them. It is therefore useful to ensure that the relatively rare occurrence of women in such roles is not the driving force in any observed relationship between female combatants and external support. I rely on the binary indicator, which simply accounts for the presence or absence of female combatants in a group, rather than the prevalence measure largely because rebel propaganda efforts do not require large numbers of female combatants. Thus, a relatively small number of female fighters may suffice to attract support from an external audience. For instance, images of female fighters are common in the propaganda materials of the PFLP, FRELIMO, and the MPLA, even though these groups employed only small numbers of women in combat operations. Given that most observers are likely to be unaware of the true extent of women's participation in a given group, reports of their participation may be sufficient to shape the perceptions of external actors. Nonetheless, clear evidence of the large-scale participation of women in the organization, as was the case with the EPLF, FMLN, and LTTE, may serve as an indication of the depth of support for the group within the base community and as a signal of its resolve to external audiences. I therefore estimate models using the prevalence indicator as a robustness check.[18]

I include a number of control variables in the analyses. First, I control for the relative strength of the rebel group. Previous studies have suggested that external actors are unlikely to extend support to groups that are particularly weak because they have a very low likelihood of victory (Gent 2008; Salehyan, Gleditsch, and Cunningham 2011). While this argument has previously been employed to explain support from external states, it likely applies to transnational advocacy networks as well. While these groups often support "underdogs," they are nonetheless likely to consider the relative strength of an actor when deciding whether or not to invest resources to support it. Extremely weak actors with a high likelihood of defeat simply represent a poor investment of the external group's finite resources. In addition, larger, more capable groups are more likely to have the capacity to engage external actors, solicit their support, and coordinate activities with them. For these reasons, I include the binary measure *much weaker*, which is coded 1 if a group was coded as such in the NSA Dataset, and 0 if otherwise.

I also account for whether the group exercised effective control over some portion of territory within the conflict state. Such control may encourage external support because it signals one type of rebel strength and facilitates rebel ability to mobilize resources away from government interference (Salehyan, Gleditsch, and Cunningham 2011). *Territorial control* is a binary indicator reflecting whether the group exercised a moderate or high degree of control over some territory outside of the government's control within the conflict state. I further include the variable *central control* to account for whether or not the group had a clear central command structure that could exercise a high degree of control over the group's forces. I include this measure because groups with more clearly defined command structures are likely to possess superior organizational capabilities and may be able to effectively utilize them to build relationships with transnational actors. Both measures come from the NSA Dataset.

I also control for the ideological orientation of the rebel group under observation. I include variables reflecting the ideologies used in previous analyses because transnational constituencies are often organized along ideological lines. For example, diaspora communities and kinship groups located in neighboring states are among the most common type of transnational constituency. Given that diaspora communities predominantly support the autonomy aspirations (or other goals) of their kinship group, I control for whether the group espoused a *nationalist* ideology. It is worth noting that transnational organizations focused on ending colonialism have also supported armed resistance movements that espoused a nationalist ideology. For instance, the ANC, SWAPO, FRELIMO, ZANU, and numerous other nationalist movements benefited from support provided by anticolonial solidarity movements located in the United States, Europe, and elsewhere. Indeed, nearly 60 percent of rebel groups in the sample that received support from transnational constituencies espoused some form of nationalist ideology. Global religious movements also represent a significant number of contemporary transnational constituencies (roughly 40 percent). For instance, transnational Islamist movements have often provided verbal support as well as material assistance to Islamist insurgencies around the globe. Somewhat surprisingly perhaps, the data suggest that leftist movements rarely receive support from a transnational constituency, accounting for only about 8 percent of the groups receiving such support. I therefore also include the control variables *leftist* and *religious* ideologies in the analyses.[19]

I also control for multiple state-level characteristics that potentially facilitate or impede rebels' ability to secure foreign support. I include the variable *democracy* because transnational advocacy groups, especially those promoting democracy and human rights, may be reluctant to extend support to armed movements rebelling against democratically elected governments, even if they agree with some of their goals or share a common ideology. I thus anticipate that rebel movements located in democratic countries will be less successful in securing transnational support. As noted above, this binary indicator comes from the "Democracy and Dictatorship" dataset.

Finally, I include controls for development level and population size. The variable *GDPpc* reflects the natural log of the average per capita GDP of the conflict country. I include this measure because rebel movements located in wealthier countries may have greater access to technologies that allow them to more easily engage external communities. For instance, communication, travel, and the transfer of funds are often easier between individuals or groups located in more-developed countries compared to those in less-developed countries. I include *population size* because large populations may reduce the external visibility of individual groups. In a large country with many competing groups and causes vying for international recognition, it may be difficult for outside observers to effectively connect with and champion the cause of any one group. I therefore anticipate a negative relationship between population size and external support.

Due to the binary nature of the dependent variables, I employ probit models with standard errors clustered by country to account for error correlation across observations within the same country. Results from these regression analyses are reported in table 5.3. Models 13 through 15 report the results of models in which the dependent variable includes both explicit and tacit support from a transnational constituency, while Models 16 through 18 report results of models that include only explicit support in the dependent variable. Model 19 presents the results of the analysis using only the subset of nationalist rebellions and the measure of diaspora support as the dependent variable. Across all of the model specifications, the measures of female combatant presence or prevalence are positively signed and statistically significant. Notably, the coefficient for *female combatants* obtained from Model 15, which uses the version of the measure that excludes cases in which women were employed only as suicide bombers, is actually larger than the coefficient in Model 14. Consequently, it appears that the presence of women in more traditional combat roles exerts a substantively larger effect. Overall, however, these results suggest that groups that contain female combatants are more likely to garner the support of transnational actors. As such, these results support Hypothesis 6. As I discussed in chapter 2, rebel propaganda is often targeted at exactly such audiences, including solidary networks abroad and co-ethnic diaspora communities. The intention of these materials is to gain sympathy from external

Table 5.3 Effect of Female Combatants on Transnational Constituency Support

	Transnational constituency			Transnational constituency (explicit)			Diaspora support
	Model 13	Model 14	Model 15	Model 16	Model 17	Model 18	Model 19
Female combatants (best)	0.621 (0.189)★★	0.533 (0.199)★★		0.538 (0.184)★★	0.408 (0.192)★		1.483 (0.755)★
Female combatants (excluding suicide bombers)			0.555 (0.208)★★			0.399 (0.210)★	
Nationalist	0.727 (0.229)★★	0.733 (0.220)★★	0.734 (0.218)★★	0.584 (0.229)★★	0.511 (0.246)★	0.509 (0.243)★	
Leftist	−0.395 (0.254)	−0.458 (0.280)	−0.496 (0.291)★	−0.211 (0.314)	−0.342 (0.358)	−0.362 (0.367)	−0.222 (0.398)
Religious	0.825 (0.257)★★	0.846 (0.260)★★	0.936 (0.255)★★	0.627 (0.247)★★	0.511 (0.265)★	0.585 (0.252)★★	−0.257 (0.639)
Much weaker	−0.410 (0.261)	−0.442 (0.272)	−0.453 (0.280)	−0.151 (0.291)	−0.314 (0.285)	−0.321 (0.292)	−1.078 (0.467)★

Territorial control	−0.069 (0.108)	−0.053 (0.106)	−0.058 (0.105)	0.082 (0.107)	0.105 (0.109)	0.102 (0.108)	0.115 (0.162)
Central control	−0.012 (0.111)	−0.009 (0.108)	−0.004 (0.109)	0.077 (0.132)	0.087 (0.136)	0.093 (0.137)	−0.289 (0.205)
Democracy		0.122 (0.215)	0.124 (0.220)		−0.160 (0.239)	−0.161 (0.242)	0.901 (0.632)
Population size†		−0.128 (0.067)*	−0.128 (0.070)*		0.006 (0.069)	0.005 (0.070)	0.224 (0.129)*
GDPpc†		0.236 (0.131)*	0.262 (0.132)*		0.356 (0.130)**	0.379 (0.131)**	0.111 (0.282)
Wald X^2	47.05	64.18	66.09	33.65	58.64	53.18	14.80
N (groups)	325.39	329.24	329.84	261.51	267.67	268.42	89.01
BIC	250	248	248	250	248	248	111

**$p < 0.01$
*$p < 0.05$ (one-tailed test)
†natural log
Note: Coefficients from probit models with standard errors (clustered on conflict country) in parentheses.

audiences, including sympathetic organizations and their members overseas, in the hopes of acquiring material or financial resources or lobbying and activism on their behalf. While fairly cursory, these results are largely consistent with those arguments.

Figure 5.4 presents the predicted probabilities of observing support from a transnational constituency. The left panel shows the predictions based on the results in Model 15, while the right panel shows the prediction from Model 18. According to the predictions, a rebel group with no female combatants in their ranks has roughly a 25 percent chance of receiving some form of support from a transnational nonstate actor. However, this probability increases to just over 45 percent for groups that include female combatants—thus, the presence of female combatants increases the chance that a group will secure external support by approximately 20 percent. The substantive effect of female combatants on obtaining explicit support, which includes material support such as funding and supplies, is smaller, but nonetheless apparent. Groups with no evidence of female combatants have about a 12 percent of securing explicit external support, while those with

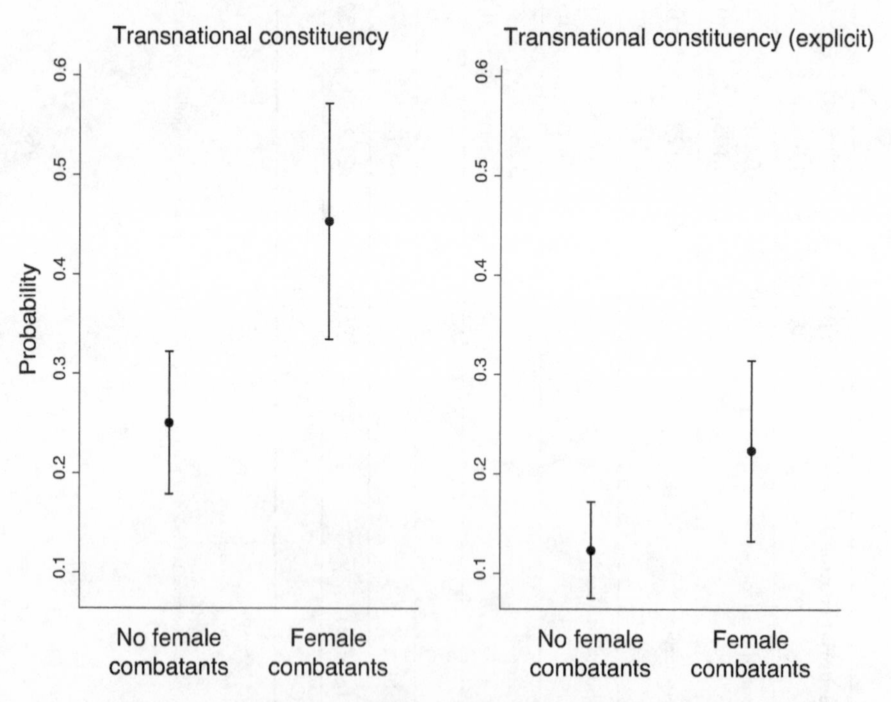

Figure 5.4 Predicted effect of female combatants on transnational constituency support

female combatants have a roughly 22 percent chance of acquiring this support. Thus, the substantive effect of female combatants on explicit support is about half of what it is for any form of support. While these results are preliminary, and should be replicated using more detailed data on the sources of support, they nonetheless suggest that female combatants exert a substantial impact on the group's success in mobilizing transnational support.

While the results are consistent with the hypotheses and robust across a variety of model specifications, it would be difficult to argue that the presence of female combatants is the primary reason that diaspora communities or transnational solidarity networks begin supporting a rebel movement. The conditions that fundamentally determine the onset or existence of such support are more likely related to the strength of existing networks between the rebel group and those communities, the ideological interests of the actors, and their strategic objectives. However, these relationships have generally been assessed only in a descriptive and theoretical sense or through studies of specific cases (e.g., Bob 2005; Nepstad 2004). Additional research is therefore necessary to identify the factors that systematically influence the likelihood of transnational actor support for rebel movements. The arguments outlined in chapter 2 and the results presented herein identify one factor that potentially increases the likelihood that these groups extend and maintain their support. By casting the movement in a more sympathetic light and making their goals appear more legitimate, female combatants may, at the margins, influence the extent to which these networks voice support for armed resistance movements.

Before turning to the discussion of the control variables, it is useful to address one additional issue related to the relationship between female combatants and external support. The argument about rebel ideology presented in chapter 1 suggests that groups espousing fundamentalist religious ideologies and those that rely on constituencies that embrace traditional gender norms are less likely to recruit female combatants, even when this strategy might allow them to ameliorate resource demands. A similar relationship is likely to obtain with respect to the role that the visible presence of female combatants plays in helping rebel groups secure external support. Specifically, while images of women in arms and the perceptions of sacrifice associated with them might encourage support and sympathy from many audiences, these images would not necessarily be expected to prompt positive responses from audiences that view the presence of female fighters as antithetical to their worldview. Just as traditionalist domestic constituencies might rebuff groups that recruit women in large numbers, transnational actors that hold deeply conservative or fundamentalist beliefs would be unlikely to support foreign rebel groups that highlight women's participation in the ranks.

To assess this corollary hypothesis, I rerun Model 15 above but include an interaction term composed of the female combatant measure and the religious ideology indicator. The expectation is that the effect of female combatants on transnational support is limited to groups that do not espouse religious ideologies. There should be no effect—or possibly a negative effect—for fundamentalist religious rebels. The results of these additional analyses are consistent with this expectation. While the interaction term is negative and insignificant, plotting the marginal effects of the interactions from these models reveals that the effect holds only in the case of nonreligious rebel organizations. As demonstrated in figure 5.5, for rebel groups that do not embrace fundamentalist religious ideologies, the presence of female combatants produces a statistically significant 20 percent increase in the probability that they will secure external support. By contrast, religious fundamentalist rebels receive virtually no additional benefits in securing support—the extremely wide confidence intervals around the coefficient

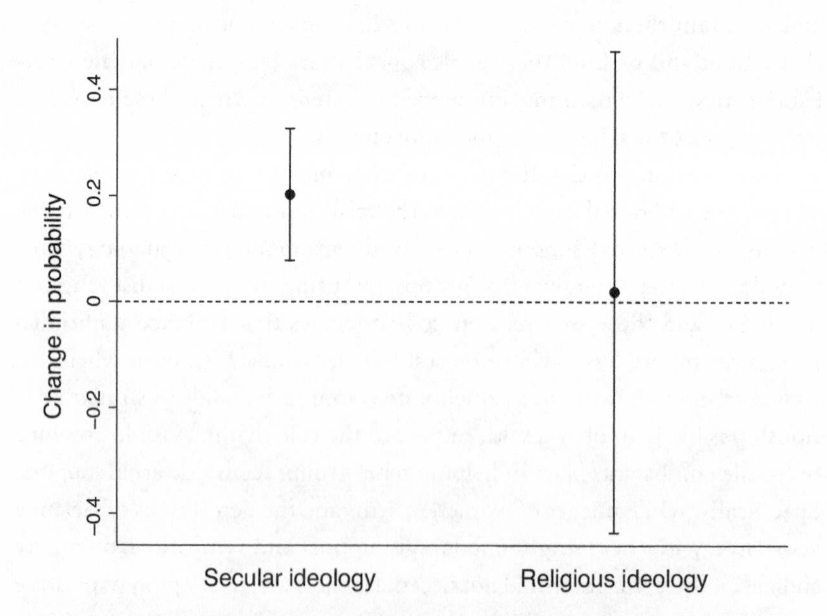

Figure 5.5 Marginal effect of female combatants on transnational constituency support by rebel ideology

estimate indicate that the effect is not significant. This result comports with the relationships identified in chapter 1 and provides additional evidence of the differential effects of female fighters in religious and nonreligious rebellions.

Several of the control variables exert a consistently significant effect on the odds of external support; however, others appear to have a limited influence. First, the coefficient on *much weaker* is negative and statistically insignificant across most of the models. Thus, while the relative strength of a rebel group has previously been shown to predict support from foreign states, it does not appear to influence the odds that it receives support from transnational actors. *Territorial control* also fails to achieve significance in any of the specifications. This result is somewhat surprising, given that the ability to control territory should facilitate a group's ability to build and maintain relationships with external actors. *Central control* also appears to have no statistically significant influence on the existence of transnational support for a rebellion. This might imply that external support is more a function of external actors' basic impressions of the group and the conflict than the result of rebel leaders' organizational capabilities and ability to cultivate networks outside of the conflict state. *Democracy* likewise fails to achieve statistical significance in the models and switches signs across the specifications. Thus, while rebel organizations may enjoy some benefits from the greater civil and political freedoms associated with more democratic states, these freedoms do not appear to enhance groups' ability to solicit external support. Neither, however, does it appear that the presence of a democratic government significantly deters external nonstate actors from offering support for rebels.

While several other group-level predictors are insignificant, the ideological variables included in the model appear to exert significant and substantively meaningful effects on the odds that a group secures support. The results suggest that movements that adopt *religious* or *nationalist* ideologies are particularly likely to again support from a transnational nonstate actor compared to other movements.[20] Global religious movements, particularly global Islamist movements in recent years, and diaspora communities are arguably more likely to support the struggles of their co-nationals or religious brethren abroad compared to other groups. *Leftist* groups, however, are apparently less likely to receive such support, though the variable only attains significance in a single specification. It is possible that global Marxism exerts a weaker influence on transnational activist support because leftist groups were already heavily predisposed to receive support from communist and socialist states, at least during the Cold War. A cursory review of major solidarity movements in the United States and Europe demonstrates that there were some efforts to support the struggles of leftist movements such as the FMLN. However, these efforts were dwarfed by the level of support provided to predominantly nationalist anticolonial movements such as ZANU, SWAPO, and the ANC.

Lastly, the results also suggest that rebels in wealthier states may be more likely to receive support from transnational actors than those engaged in conflicts in less wealthy states, while states with larger populations may be less likely to receive such support. *GDPpc* is positive and significant in four out of the five models in which it is included. This is perhaps because wealth facilitates communication and mobilization between activists and the conflict state and those located abroad. Finally, the effect of *population size* varies substantially across the models. While it is negative and achieves statistical significance in the models predicting any transnational support, it becomes positive and insignificant in the models predicting only explicit support. It then becomes positive and significant in the model using diaspora support as the dependent variable. It is difficult to explain what drives such changes, but in at least some contexts the variable appears to be a relevant predictor of constituency support.

Discussion and Conclusion

In chapter 2, I argued that recruiting female fighters and highlighting their presence and participation in the group benefits rebels in a variety of ways. First, it expands the supply of available recruits, allowing rebels to field larger, more capable military forces. Second, it helps rebels build sympathy and legitimacy in the eyes of external audiences, leading to increased support for the group and its goals. These factors should in turn increase the likelihood that the group receives support from external nonstate actors (e.g., diaspora communities or transnational organizations). While these benefits may be more likely to accrue to rebel groups that include female fighters, I do not argue that securing foreign support is a primary motive for the initial decision by rebels to recruit women and deploy them in combat. Rather, as I argued in chapter 1, a mixture of ideology and resource demands are probably the most influential factors in that decision. Nonetheless, once the group has incorporated women into its fighting force, its leaders often recognize the potential role the image of female combatants can play in garnering sympathy and support for the movement. Many rebel groups therefore strategically highlight female fighters in their propaganda and recruitment materials as a means of attracting external (as well as domestic) support.

The results presented above provide support for these arguments. First, I find that rebel groups that include larger numbers of female combatants are also more likely to field a larger number of troops compared to groups that exclude female combatants or recruit them in only very small numbers. Second, I find evidence that the presence of female combatants positively influences audience attitudes toward the groups that employ them. The results from a survey experiment designed to probe this relationship suggest that simply being made aware of the presence of female combatants in the groups improved respondent support for the group and its goals. Third, I argued that, given the influence of female combatants on observer attitudes, rebel movements that include female fighters should be more likely to attract support from transnational solidarity organizations and diaspora communities than those that exclude them. The results of multiple quantitative analyses support the arguments regarding the relationship between female combatants and external support for the rebel group. Specifically, they provide empirical evidence that transnational nonstate actors such as diaspora communities and international solidarity networks are more likely to extend support to rebel groups in which female fighters actively participate.

As a final note, it is important to reiterate that expanding the supply of available troops and acquiring foreign support are not end goals for these groups. Rather, successfully achieving these objectives is valuable to rebels because they shape the trajectory of the conflict and influence the group's likelihood of survival and the odds that it will eventually achieve broader political and military goals. Thus, to the extent that female fighters assist the group in achieving these lower-order objectives, they should indirectly influence the group's ability to survive and achieve its broader objectives. In other words, the argument and results imply that not only do female fighters provide specific strategic benefits such as a larger supply of troops and increase likelihood that the group receives external support, they may also indirectly influence the outcome of the conflict.

Conclusion

Understanding Women's Participation in Armed Resistance

Until recently, scholars of political violence have largely overlooked women's participation in armed conflict, and many still consider it a peripheral area of inquiry. In part, this view is based on the assumption that men—and almost exclusively men—fight wars, and women, when they do appear in the ranks of armed groups, are simply novelties or outliers. This narrow perspective, however, ignores the important contributions that women have made to the various rebel movements for which they have fought (and often died). Moreover, this view fails to appreciate the extent to which the recruitment of female fighters contravenes embedded social norms, and it thus unintentionally minimizes the gravity of this decision for both the rebel leadership and the women that take up arms. One observable consequence for the construction of social science theory is that while civil conflict scholars have developed a sophisticated understanding of the mechanisms and processes driving civil war onset, rebel resource mobilization, conflict outcomes, and other aspects of internal conflict dynamics, issues such as women's contributions to civil conflicts, the factors motivating their entry into armed groups, and the implications of their presence in these organizations—and for the societies to which they will ultimately return once the violence subsides—remain poorly understood.

Fortunately, however, scholars have increasingly devoted attention to the phenomenon of women's participation in armed nonstate political organizations in recent years. Detailed qualitative investigations of specific cases have highlighted the pathways though which women enter armed groups; assessed the motives for their mobilization, which are often very similar to those of male combatants; and identified the social conditions and political dynamics that help explain women's participation in specific armed resistance movements (e.g., Coulter 2009; Hamilton 2007; Kampwirth 2002; Viterna 2013). An emerging body of quantitative research has also sought to systematically assess the impact of organizational characteristics and the features of the conflict environment on the likelihood that armed groups employ women in combat roles (e.g., Henshaw 2017; Thomas and Bond 2015; Wood and Thomas 2017). Despite the important findings generated by these studies, many questions regarding both the decisions of armed groups to accept women into their ranks and the decisions of the women who pursue these opportunities remain unanswered.

In this book, I have chosen to address questions related to women's participation in armed conflict primarily from the standpoint of the leaders that recruit them. In so doing, my intention has been to help clarify the conditions under which rebel movements are more or less likely to extend combat roles to women and to put forth a generalizable argument that accounts for the wide variation in the prevalence of female fighters across contemporary armed rebellions. In addition, I have attempted to offer some initial insight into the potential strategic implications of this decision for the groups that employ women combatants. In this concluding chapter, I first review the main empirical findings from the previous chapters and then offer a small number of important caveats and qualifications to the preceding arguments and findings. I include this brief discussion because I believe it is important to clearly explicate the scope of the findings and to highlight not only the ways in which they shed light on the questions at hand but also note the aspects of the question that remain unaddressed or underaddressed. Finally, I offer a few brief comments on the theoretical implications of the arguments, note several potentially fruitful related areas of future inquiry, and briefly discuss a few of the relevant policy implications of the findings.

Major Findings

As I discussed in the introduction and at various other points throughout the book, men (and masculinity) have historically been associated with violence and warfare, and women (and femininity) have traditionally been associated with innocence, virtue, and peace. The strength of these associations varies both nationally and by specific community. However, the wartime division of labor, in which men fight to defend the country (or community) and women represent both the cause for which men risk their lives as well as their reward for doing so, is present across diverse cultures and persists overt time. During periods of violent armed conflict, men therefore often face intense social pressures to conform to traditional masculine roles and participate in combat, while women are discouraged from undertaking such activities and are instead expected to perform more traditionally feminine roles such as providing care, support, and encouragement to male fighters.

One of the implications of these embedded norms and attitudes is that the presence of armed young women represents a striking deviation from the historical norm. While female fighters are more common among rebel, insurgent, terrorist, and other armed resistance movements compared to state militaries, they remain a small proportion of the total number of combatants in virtually all contemporary conflicts. Nonetheless, there exists substantial variation across armed resistance movements in terms of the roles women play and the proportion of female combatants in their ranks. The MFDC in Senegal and ASG in the Philippines have, for example, never employed women in combat, and appear to have few if any formal female members at all. PIJ in Palestine/Israel has recruited very small numbers of women to serve as suicide bombers but has largely excluded them from other military roles. The PFLP, another Palestinian organization, also employed only a small number of female combatants, and it typically only utilized them in terror operations such as airline hijackings, sabotage, and assassinations. Yet, many women were included among their formal members, including many who actively participated in the group's armed wing in noncombat support roles. By contrast, female fighters made up a substantial portion of the combat forces of the FMLN in El Salvador, TPLF in Ethiopia, LTTE in Sri Lanka, and the RUF in Sierra Leone. In each of these rebel groups, women participated in both noncombat and combat roles, and some women (albeit only a small number) attained leadership positions.

What explains this variation? Why do some movements completely exclude women from the battlefield or relegate the small numbers of women they recruit to background and support roles, while others integrate large numbers of female fighters into their armed wings? These represent the motivating questions for this project and the primary questions addressed in chapter 1. In order to answer these questions, I adopted a framework that focused on demand-side factors in rebel resource mobilization. Using this approach, I identified the severity of the resource demands imposed on rebel movements and their sensitivity to the perceived costs associated with the mobilization of women for war as two important explanatory factors in determining the presence and prevalence of female combatants in an armed group. In brief, I argued that rising resource demands, such as those that occur during periods of rapid conflict escalation and the inability to locate sufficient numbers of male volunteers, create strong incentives for the leadership to adopt new recruitment strategies. Along with forcible recruitment and child soldiers, the recruitment of female combatants represents one of several alternative mobilization strategies available to rebel leaders. All else being equal, rising resource demands increase the prevalence of female combatants in an armed movement.

However, as I argued, all is not equal across rebel groups. Whether or not rebel leaders elect to incorporate women into their fighting forces when they face such demands depends on their ideological orientation and, relatedly, the beliefs and norms of the constituent communities on which the group relies for support. While rebel groups espousing nonreligious ideologies become increasingly likely to recruit female combatants as conflict costs increase, fundamentalist and orthodox religious movements are comparatively less susceptible to these pressures. Rather, they experience countervailing pressures that disincentivize the use of female combatants as a means of ameliorating acute resource demand. Even as resource demands mount, leaders of rebellions espousing fundamentalist or orthodox religious ideologies generally refrain from recruiting female combatants. And even when they do elect to recruit women, they are likely to do so only in small numbers and to deploy them in only very limited roles (e.g., as suicide bombers). These arguments imply that for many rebel groups, the decision to recruit female combatants is a strategic one. Rebel leaders typically recruit women in large numbers and deploy them in combat roles not because of any deep commitment to gender equality, but because they believe that this strategy will improve their ability to achieve their goals, or at least promote the group's survival. Left-wing ideology and gender egalitarian beliefs serve as moderating factors in this decision.

The recognition that female recruitment is often a strategic decision prompts the discussion in chapter 2. Therein, I considered the strategic implications of the decision to recruit female fighters. I made two main arguments in that chapter. First, I argued that the use of female fighters ultimately helps rebel groups ameliorate their human resource constraints and increases the size of their combat forces, which is often the impetus for their recruitment. Second, I argued that the visible presence of female fighters strategically benefits the group by drawing attention to its cause and by improving domestic and external audiences' attitudes toward the group. Moreover, I argued that highlighting women's participation in the movement represents one strategy though which rebel groups can increase their likelihood of attracting support from transnational solidarity networks and diaspora communities.

I began the empirical investigation of these hypotheses with a set of case illustrations. While these vignettes do not represent explicit tests of the hypotheses, they nonetheless demonstrate the validity of the central arguments. Closer scrutiny of the PKK and its allied militias in Turkey and Syria, the LTTE in Sri Lanka, and ZANLA and ZIPRA in Zimbabwe illustrates the relationship between the rising resource pressures generated from the intensification of the conflict and leaders' efforts to recruit women and incorporate them into the groups' fighting forces. In each case, the prevalence of female combatants increased as the conflict intensified; moreover, the available evidence suggested that the leaderships' made their decisions to expand (or initiate) women's recruitment during periods of rapidly rising conflict intensity and sharply rising resource demands.

The case illustrations also highlight the roles that group ideology played in the evolution of women's participation. As an avowedly Marxist movement, the PKK was minimally constrained by the prevailing social norms of the conservative Kurdish society on which it relied for support. Nonetheless, there appeared to be a coevolution in the group's rhetoric of gender egalitarianism and their efforts to recruit female combatants. This suggests the possibility that the increase in gender egalitarian rhetoric was at least partly instrumental. The recruitment of women into the LTTE followed a somewhat similar trajectory, though the group's adherence to prevailing social values was initially comparatively stronger and the extent of its change was more dramatic. While the LTTE did not embrace Marxist ideology the group's use of gender egalitarian rhetoric increased as it sought to bring more women into the movement and when it faced increasing demands for troops. Similarly, while lacking any deep commitment to Marxism, both ZANLA and ZIPRA espoused a nominally socialist political platform. Both also employed the language of women's rights and emancipation in their rhetoric. However, there is some evidence that this decision was driven as much by strategic interests as by ideological commitment, and while they eventually deployed women in combat roles, this decision appeared to come later than in the other movements, and only after a steep escalation in the intensity of the conflict.

Chapter 3 also illustrates the manner in which each rebel group—and sometimes their allies abroad—attempted to highlight women's participation as a part of their propaganda and recruitment efforts. The PKK's leadership recognized the recruitment benefits of female fighters; however, there is little evidence that they used female combatants in their international propaganda efforts until recent decades. Since the late 2000s, the PKK and sympathetic transnational organizations have explicitly highlighted the high prevalence of female combatants in its ranks to garner attention and sympathy for the group and its cause. By contrast, the LTTE recognized the benefits associated with drawing attention to women's participation in the violence quite early. It produced numerous videos and publications that showcased female cadres' dedication and sacrifices in the conflict. The role of female fighters in the propaganda efforts of ZANLA and ZIPRA is somewhat less clear. Nonetheless, it appears that both groups (particularly ZANLA) intentionally highlighted women's roles as fighters and their inclusion in leadership positions (even if modest in reality) to establish their legitimacy and attract external support during the later stages of the conflict.

Chapters 4 and 5 were devoted to empirically examining the hypotheses drawn from the previous chapters. In chapter 4, I examined the empirical implications of the theory presented in chapter 1. The statistical results provide robust support for the arguments that resource demands and group ideology jointly influence the prevalence of female combatants in an armed group. Moreover, I found evidence that the general relationship between rising resource demands and the decision to recruit female combatants is conditioned by the group's ideological orientation. While increasing resource demands result in an increase in female combatants among rebel groups with nonreligious ideologies, such demands have no discernible impact on the likelihood that groups espousing religious ideologies recruit female combatants. In other words, orthodox and fundamentalist rebel groups persist in excluding female combatants even as conflict costs mount and recruitment become more difficult.

The results of various quantitative analyses—using both observational and experimental data—presented in chapter 5 provide empirical support for the arguments presented in chapter 2. First, they suggest that rebel groups that decide to recruit female combatants ultimately field larger combat forces than those that exclude female combatants or recruit them in only very small numbers. Second, results from a survey experiment provide some evidence, albeit preliminary, that the visible presence of female combatants in an armed group can positively influence observer impressions of the movement and its goals. In the experiment, respondents expressed more positive sentiments toward a hypothetical rebel group and its goals when they were informed that the group included female combatants in its ranks and were provided an image that subtly but clearly highlighted their presence. The substantive impact of this effect was modest, but the result nonetheless suggests that awareness of women's inclusion as fighters within an armed resistance movement can subtly shape the opinions of outside observers toward the group and the conflict. Lastly, I find support for the claim that rebel groups that include female fighters are more likely to receive support from transnational nonstate actors such as diaspora communities and international solidarity networks.

Caveats, Conditions, and Extensions

The arguments presented in this book as well as the empirical results help identify the conditions under which female combatants are more or less likely to emerge during armed internal conflicts and the ways in which groups capitalize on their presence for strategic gains. However, these findings should not be interpreted as the definitive explanation of female combatant recruitment. Nor are they intended as a complete accounting of the ways in which female combatants influence the groups for whom they fight or the way these groups interact with other actors in their strategic environment. While I identify two important general factors that shape rebel leaders' decisions to create opportunities for women, it is important to recognize the limitations of these arguments and the roles that additional factors might play in determining or conditioning rebel groups' decisions to utilize female combatants.

First, local conditions likely moderate the relationships articulated in this book. I address only one such factor here, and I do so only indirectly. Specifically, I assume that the constituent community's commitment to traditional gender norms and hierarchies is likely to influence rebel leaders' willingness to incorporate large numbers of women into the organization. I further assume that a group's ideology determines its perceptions of the potential costs generated by transgressing these norms, including any backlash from this constituency. While acknowledging the role of local norms in conditioning rebel decision-making with respect to the use of female fighters, the argument does not fully account for these important dynamics. Understanding the interaction between a rebel group's willingness to recruit female combatants and the underlying social constraints embedded in the local environment requires examining specific cases where the effects of local economic, social, and political factors can be theorized in a more nuanced manner. The arguments presented here should therefore be viewed as a starting point for understanding the ways in which rebel leaders balance the (often) countervailing demands of resource pressures and constituency attitudes—what I term "the female recruitment dilemma."

Second, in constructing the primary arguments put forth in this book, I prioritize the decision-making of the leadership over the decision calculi of individual recruits. These arguments therefore pay only limited attention to the factors that influence women's decisions to join armed groups. Women do not constitute a homogenous or monolithic group, and different subgroups within the female population are likely to respond to different recruitment stimuli. Only a handful of previous studies have systematically examined patterns of recruitment within the female population or the diverse factors that motivate or deter women from joining armed groups when the opportunity arises. Consequently, there is little understanding of why some women readily choose to join armed groups when the option is made available, while other women remain uninvolved even when they support the group's social or political goals.

The scant existing research in this area suggests that factors such as prior experience with political activism, the presence of network ties to the organization, presence in refugee camps, and personal biographical characteristics (especially having children) influence women's decisions to engage in high-risk collective action, including joining an armed rebellion (e.g., Viterna 2006, 2013; Parkinson 2013). The nature of the theoretical framework and data used in this study effectively prohibit any deeper investigation of the factors that incentivize women (as a group) or specific subgroups within the female population to support or join a rebel movement. In other words, not only can we infer little about the role of supply in determining the prevalence of female combatants in an organization, we cannot make valid inferences about the conditions that shape the supply in the first place. Additional research is needed in this area.

Third, I do not presume that the presence or absence of female combatants in an armed movement is the pivotal factor determining whether or not a group receives support from domestic or foreign actors or the level of sympathy it garners from foreign or domestic audiences. That would simply be naïve. The presence of female fighters is neither a necessary nor sufficient condition for acquiring such support. Rather, strategic, ideological, and logistical factors are likely to have the greatest influence on this decision. Nonetheless, as I discussed in chapter 2, by shaping audience beliefs about the movement and its goals, the presence of female fighters can exert a nominal influence on the odds that the group receives external support. The gender composition of a rebel group is simply one characteristic of a movement, but it is a potentially influential characteristic that previous studies have largely ignored.

Moreover, as with the motivation to recruit female combatants, a number of factors are likely to condition the relationship between female combatants and external support. Particularly, the gendered beliefs and attitudes of the audiences to which the group broadcasts its appeals for support likely influence whether and to what extent the visible presence of female fighters improves observers' opinions of the group and its goals. I made a general claim informed by the existing literature regarding gendered imagery and war that images of armed young women risking their lives for a political movement "tugs on the heartstrings" of observers and helps legitimize the group. However, it is possible that the same image might offend some audiences and ultimately delegitimize the group. A supplementary analysis presented in chapter 5 provides some evidence for this contention by showing that the benefits of female combatants in garnering external support extend only to secular rebel groups. Future analyses that investigate the role of female fighters in rebel propaganda and recruitment efforts should carefully consider how assumptions about these groups' intended audiences and their values lead rebel movements to construct their propaganda messages, particularly with respect to female combatants and other gendered aspects of conflict.

Theoretical Implications

The findings in this book have a number of potential implications for existing areas of research in political violence and conflict processes. First, these findings should have clear implications for research related to the duration and outcome of civil conflicts. A central argument put forth in the book is that rebel groups turn to the mobilization and deployment of female combatants in order to stave off defeat as the conflict intensifies and resource pressures mount. I also argue that, on average, this strategy is successful, and rebel groups that recruit female combatants tend to mobilize larger combat forces than those that do not. Moreover, while I do not assume that the effort to secure foreign support motivates female recruitment, I do find evidence that the presence of female combatants can assist groups in achieving this objective. As noted in chapter 2, previous studies generally suggest that groups that can more effectively mobilize resources are less likely to suffer defeat and are more likely to achieve their goals. To the extent that incorporating female combatants into a group's fighting force both ameliorates resource demands and (subsequently) increases the odds that it can secure support from foreign actors, the presence of female combatants may ultimately influence the duration of the conflict and the likelihood that the group achieves its goals.

These findings may also help inform the future construction of theories on rebel resource mobilization and recruitment more broadly. As discussed in chapter 1, rebels choose among a variety of different resource mobilization strategies (forcible recruitment, child soldiers, female combatants, foreign fighters, and so on). Each strategy carries with it different potential benefits and different potential complications and costs. In addition, these strategies are not mutually exclusive, and rebels have the option of simultaneously combining multiple mobilization strategies or substituting one strategy for another over time if they choose. Scholars have typically studied these strategies in isolation; however, a more general theory about how rebels select their resource mobilization strategies would represent a major advancement. The arguments presented in chapter 1 might help propel such an endeavor. The supply-demand model of rebel recruitment has now been applied to a variety of mobilization strategies, including foreign fighters (Hegghammer 2013), child soldiering (Achvarina and Reich 2006; Andvig and Gates 2010), forcible recruitment (Eck 2014), and female combatants (Thomas and Bond 2015; Wood and Thomas 2017). Future analyses could refine this basic model in order to better understand how rebel leaders choose among these available strategies or the conditions under which they use them as substitutes or complements for one another. Such a theory likely requires more detailed consideration of the intervening variables and mitigating circumstances that facilitate or constrain the use of these strategies.

The observation that women are often willing to mobilize to support rebel movements when they are provided with an opportunity represents another potential implication for theories of rebel resource mobilization. Most existing theories of popular mobilization for armed conflict implicitly assume that the population of interest is male. This gendered assumption may therefore shape scholars' assessment of the mechanisms through which political entrepreneurs successfully solve the collective action problem or their appraisal of the effectiveness of various resource mobilization strategies. We might reasonably infer that women, like men, weigh the costs and benefits associated with high-risk collective action and make a decision based on the balance of those factors. Yet, is it reasonable to assume that men and women endure similar costs for joining rebellions? Do they perceive the same rewards from participation? If they do, then existing theories are sufficiently generalizable. If we can assume that these costs and rewards differ but that men constitute the entirety of the population of interest, existing theories still suffice. Yet, if neither assumption holds—implying that there are gender-specific cost and reward functions and that women represent a substantial part of the population rebel leaders wish to mobilize—then existing theories could be improved by accounting for these facts.

The role of ideology as a mobilization force is perhaps instructive. Ideological commitment has long been viewed as an important factor in recruitment and mobilization (e.g., Gates 2002; Weinstein 2007). Yet, similar ideological convictions might lead to different decisions by men and women. For example, while both devout men and women committed to orthodox religious ideologies are likely to support the goals of armed groups that share their convictions, male and female responses to offers of recruitment by such groups are likely to substantially differ. As Reed Wood and Jakana Thomas (2017) assert, while small numbers of women who embrace orthodox religious ideologies are willing to transgress social and religious norms to serve as combatants, the majority of these women are likely to view direct participation in organized violence as anathema to their beliefs about women's appropriate roles in society. As such, they are likely to locate other avenues through which to help the group advance its goals, including providing comfort and encouragement to male fighters, instilling their ideology in their children, and fund-raising and propagandizing for the group. Additional research into the factors that shape the supply of female recruits is essential to developing a more complete theory of women's participation in armed resistance movements.

The arguments and results in this book also have direct and indirect policy implications. First, increased recognition of the scope of women's participation in armed conflict, both as active combatants and in support roles, can hopefully increase policymakers' awareness of women's presence and contribute to larger and more focused efforts to reintegrate women combatants and women associated with armed groups into post-conflict societies. Efforts to address the needs of these women have increased in recent decades. However, the international community often underestimates or overlooks women's participation, and relatively few DDR processes contain specific provisions for female combatants. Of particular concern, in many cases only frontline combatants (or those who possessed weapons) are included in such programs. Yet, as highlighted in this book, women's participation in combat is often situation dependent, and many female combatants only participate in combat on an as-needed basis. Similarly, rebel commanders often quickly transition women out of the group's combat forces as the conflict de-escalates. Women who participated in warfare may therefore be excluded from the DDR process, inhibiting their ability to successfully reintegrate into society and creating long-lasting negative psychological and social repercussions. Women who participated in conflict often face substantial challenges to reintegration due to both the psychological trauma of war and the lingering social stigma associated with violating traditional norms by taking up arms (see Bouta, Ferks, and Bannon 2005; El-Bushra 2003; McKay 2004). They are, for example, less likely to marry and less likely to re-engage with their prewar networks. They also often face greater economic vulnerability and experience higher rates of domestic abuse than women who did not participate in armed conflict.

Finally, by illuminating the potential role that female combatants play in assisting rebel groups in mobilizing domestic and international support, the arguments and findings presented herein should be of use to policymakers seeking to understand how rebel groups interact with local communities, the sources and mechanisms of their support, and for strategies to curtail or suppress rebel recruitment. However, the recognition that women make substantial contributions to armed groups, and that those contributions may have significant effects on the conflict, might also generate externalities that further negatively impact women in conflict environments. Wartime counterinsurgency efforts often reflect a gendered understanding of political violence. Particularly, while state forces often brutalize women, young males are disproportionately targeted for violence in repressive counterinsurgency campaigns. Increased attention to the role of female combatants could ultimately contribute to an increase in the use of repression against the female population. Policymakers and human rights advocates might therefore need to pay particular attention to these dynamics in conflicts that involve large numbers of female fighters.

APPENDIX A

Version History

This appendix summarizes updates to the original dataset used in Wood and Thomas (2017). The changes detailed below reflect new information uncovered since the publication of the original dataset and additional cases included by expanding the coding back to 1964. For each case, I include the Dyad ID number from the UCDP dataset, the group name, the country in which the conflict was located, and the original and updated coding ("best"/"high").

The full dataset is available at the author's website (https://reedmwood.com).

Table A.1 Version History

UCDP ID	Name	Country	Previous code (best/high)	New code (best/high)	Notes
7	UNITA	Angola	0/3	1/3	
25	Abkhazian separatists	Georgia	0/1	1/1	
36	Al-Qaeda	United States and others	Missing	1/1	Suicide bomber only
40	RCD	Democratic Republic of the Congo	0/1	1/1	
131	AFRC	Sierra Leone	Missing	1/2	
149	Holy Spirit Movement	Uganda	Missing	1/1	
286/453	NCSN-IM/K	India	0/0	0/1	Split into distinct factions. Scores unchanged
303	Serbian Republic of Krajina	Bosnia	Missing	1/1	
306	KNU	Myanmar	2/3	0/1	
310	NMSP	Myanmar	Missing	1/3	
312	RSO	Myanmar	Missing	0/0	
343	EPL	Colombia	Missing	1/1	
364	Khmer Rouge	Cambodia	1/1	1/2	
367	PMR	Moldova	Missing	0/0	

UCDP ID	Name	Country	Previous code (best/high)	New code (best/high)	Notes
378	KPNLF	Cambodia	Missing	0/0	
382	Lao Resistance Movement	Laos	Missing	0/0	
383	Christian militias	Lebanon	2/3	1/2	
419	PFLP	Israel/Lebanon	1/1	1/2	
435	EJIM	Eritrea	Missing	0/0	
446	Ahlul Sunnah Jamaa	Nigeria	Missing	0/0	
448	ISIS	Iraq/Syria	0/0	1/1	Suicide bomber only (beginning 2018)
464*	ELN	Bolivia		0/1	
500*	ELN	Peru		0/0	
521*	Islamic Charter Front	Sudan		0/0	
544*	Baluchi insurgents	Pakistan		0/0	
548*	Muti Bahini	Bangladesh/ Pakistan		1/1	
560*	ZANU	Zimbabwe/ Rhodesia		2/3	

(*continued*)

Table A.1 Version History (*continued*)

UCDP ID	Name	Country	Previous code (best/high)	New code (best/high)	Notes
565*	ZAPU	Zimbabwe/ Rhodesia		2/2	
564*	Socialist faction	Yemen		0/0	
586*	CCO	Malaysia		0/0	
593*	Montoneros	Argentina		2/3	
594*	ERP	Argentina		2/3	
614*	NLF/Viet Cong	South Vietnam		2/2	
616*	Royalists	North Yemen		0/0	
634*	Pathet Lao	Laos		0/1	
656*	FROLINAT	Chad		0/0	
675*	PAICG	Guinea-Bissau		1/1	
676*	Anya/Nya	Sudan		0/1	
679*	CNL	Democratic Republic of the Congo		0/0	
681*	FLNC	Democratic Republic of the Congo		0/1	
685*	FRELIMO	Mozambique		1/2	
687*	EPRLF	Sri Lanka	0/0	0/1	
690*	CPI-ML	India		1/1	
711*	NNC	India		0/0	

UCDP ID	Name	Country	Previous code (best/high)	New code (best/high)	Notes
727*	PDPA	Afghanistan		0/1	
734*	Republic of Biafra	Nigeria		1/2	
738*	MPLA	Angola		1/2	
739*	FNLA	Angola		0/0	
746*	SSA	Myanmar		0/1	
754*	EDU	Ethiopia		0/0	
775*	MNDAA	Myanmar		0/0	
790*	KNUP	Myanmar		0/0	
796*	SALF	Ethiopia		0/0	

*New cases added to dataset covering years 1964 to 1978

APPENDIX B

Examples of Coding Narratives from WARD

For each case included in WARD, we produced a short coding narrative that describes the ways in which women participated in the armed group and includes a summary of relevant bibliographic information. The following represent examples of such narratives from cases in Latin America and Africa. The full collection of such narratives is available at: https://reedmwood.com/home-page/women-in-armed-rebellion-dataset-ward/.

Contras/Democratic Forces of Nicaragua (FDN)

Female formal participants: Yes
Female combatants: Yes
Female combatant prevalence: 2/3
Female suicide bombers: No

Women occupied both combat and noncombat roles in the Contra forces. Surprisingly specific statistics were recorded regarding the involvement of women in the Contras. By one estimate, 39 percent of rebels demobilized at war's end were women, but this included relatives and other persons residing at rebel camps in Nicaragua and Honduras. Women were involved in combat, but their participation in front-line combat was limited. Overall, roughly 6 percent of all "commandos" killed were women, and there is evidence that several hundred female Contra fighters were killed in combat. Based on available estimates, female combatants appear to have composed between 7 and 15 percent of the Contra combat forces, though some figures (that likely included noncombat roles) place that figure as high as 20 percent. Numeric estimates place the total number of female combatants as high as three thousand. Although female combatants were common, they were not encouraged to attain a rank higher than or equal to that of a comandante; nonetheless, some women appear to have served in decision-making positions and attained field-level leadership positions (Brown 2001).

Sources

Brown, Timothy. 2000. *When the AK-47s Fall Silent: Revolutionaries, Guerrillas, and the Dangers of Peace.* Stanford, CA: Stanford University Press.

——. 2001. *The Real Contra War: Highlander Peasant Resistance in Nicaragua.* Norman: University of Oklahoma Press.

Gonzalez-Perez, Margaret. 2006. "Guerrilleras in Latin America: Domestic and International Roles." *Journal of Peace Research* 43 (3): 313–329.

——. 2008. *Women and Terrorism: Female Activity in Domestic and International Terror Groups.* New York: Routledge.

Kampwirth, Karen. 2001. "Women in the Armed Struggle in Nicaragua: Sandinistas and Contras Compared." In *Radical Women in Latin America: Left and Right,* ed. Victoria Gonzalez and Karen Kampwirth, 79–101. University Park: Pennsylvania State University Press.

19th of April Movement (M-19) (Colombia)

Female formal participants: Yes
Female combatants: Yes
Female combatant prevalence: 3/3
Female suicide bombers: No

The 19th of April Movement (M-19) began as an armed wing of the leftist National Popular Alliance (ANAPO), and it was mostly made up of young, middle-class students. The M-19 differed from other armed rebel groups in Colombia because it was born as an urban organization, seeking to connect with people in big cities to carry out their leftist vision rather than rural communities (Holmgren 2009).

Exact figures for women's participation in M-19 are unavailable. However, numerous sources suggest that women constituted as much as 30 percent of the M-19's forces. Female members received military training and held a variety of roles in the organization, including noncombat roles (intelligence, safe-house operation, logistics), combat positions (assassinations, terror attacks, hostage taking, sabotage, and so on), and leadership positions. Although the M-19 was more open to female leadership than other rebel organizations, its internal system was still highly patriarchal and viewed women as less capable than men. Women were generally not forcibly recruited into the M-19; rather, they were drawn to the group through political conviction and leftist ideology. Male leaders in the M-19 often relied on gender norms and used the femininity of their female fighters to the group's advantage.

Sources

Gjelsvik, Ingvild Magnæs. 2010. *Women, War, and Empowerment: A Case study of Female Ex-combatants in Colombia*. Tromsø, Norway: University of Tromsø.
Holmgren, Linda Eitrem. 2009. *The Woman Warrior: A Post-Structural Gender Analysis of Guerillas in Colombia*. New York: Lund University.
Zasquez Perdomo, Maria Eugenia. 2005. *My Life as a Colombian Revolutionary: Reflections of a Former Guerrillera*. Philadelphia: Temple University Press.

Tigrayan People's Liberation Front (TPLF) (Ethiopia)

Female formal participants: Yes
Female combatants: Yes
Female combatant prevalence: 3/3
Female suicide bombers: No

Consistent with the pattern observed in many other Marxist-Leninist rebellions, women occupied many roles in the TPLF. These included both noncombat and combat duties. Numerous sources place the proportion of female fighters in the TPLF at around 30 percent, though it should be noted that the term "fighter" may apply to women in noncombat roles as well. Nonetheless, the available evidence points to a high level of women's involvement in the armed wing of the TPLF and to a large number of trained, armed female combatants in the group's combat forces.

As in most groups, the proportion of women in the group and the roles they played varied over time. The TPLF was founded in 1975, and women were initially not welcomed into the movement. However, by 1982 approximately one-third of the fighters were women. In the mid-1980s, however, the TPLF leadership decided to restrict the number of female fighters. The reason given was that too many women were volunteering to the join the movement, and it was beginning to disrupt domestic life in the countryside. The leadership raised the educational requirement to five years of schooling, and many women were unable to meet this requirement.

Interestingly, unlike many rebel groups where women are more prevalent in local militia or auxiliary units compared to the national force, the opposite appears to be true in the TPLF. In many places women constituted less than 5 percent of these units or were nonexistent. The (partial) explanation provided for this pattern is that it was more difficult for young women to transgress gender norms and escape social expectations when they were close to home compared to when they served as fighters in the formal TPLF force and were more likely to be sent far from home.

Sources

Hammond, Jenny. 1990. "'My Revolution Is Like Honey': Women in Revolutionary Tigray." *Women: A Cultural Review* 1 (1): 56–59.

Negeow-Oda, Beza, and Aaronette M. White. 2011. "Identity Transformation and Reintegration Among Ethiopian Women War Veterans: A Feminist Analysis." *Journal of Feminist Family Therapy* 23: 161–187.

Tadesse, Medhane, and John Young. 2003. "TPLF: Reform or Decline?" *Review of African Political Economy* 97: 389–403.

Tareke, Gebru. 2009. *The Ethiopian Revolution: War in the Horn of Africa.* New Haven, CT: Yale University Press.

Veale, A. 2003. *From Child Soldier to Ex-Fighter: Female Fighters, Demobilisation and Reintegration in Ethiopia.* Pretoria, South Africa: Institute for Security Studies.

Young, John. 1997. *Peasant Revolution in Ethiopia.* Cambridge: Cambridge University Press.

Movement of Democratic Forces of Casamance (MFDC) (Senegal)

Female formal participants: No
Female combatants: No
Female combatant prevalence: 0/0
Female suicide bombers: No

The available evidence suggests that while numerous women supported the MFDC and many served as civilian activists and collaborators, there was no formal role for women in the armed wing of the organization. Despite references to women's participation in political protests in movements allied with the MFDC and to women providing food and other goods to the rebels, there is no evidence that women ever received military training or participated in combat operations. One source explicitly asserts that "there have been no reports of female combatants" (Stam 2009).

Sources

Stam, Valerie. 2009. "Women's Agency and Collective Action: Peace Politics in the Casamance." *Canadian Journal of African Studies* 43(2): 337–366.

Tolivera-Diallo, Wilmetta. 2005. "The Woman Who Was More Than a Man: Making Aline Sitoe Diatta into a National Heroine in Senegal." *Canadian Journal of African Studies* 39 (2): 338–360.

USAID. 2001. *West Africa: Civil Society Strengthening for Conflict Prevention Study. Conflict Prevention and Peace Building Case Study: The Casamance Conflict and Peace Process (1982–2001).* https://gsdrc.org/document-library/civil-society-strengthening-for-conflict-prevention-study-conflict-prevention-and-peace-building-the-casamance-conflict-and-peace-process-1982-2001/.

APPENDIX C

Survey Wording and Instrument

This appendix provides additional details of the survey experiment described in chapter 5. The survey instrument was designed in collaboration with Dr. Devorah Manekin (Hebrew University of Jerusalem) and was deployed to a sample of U.S. respondents organized by Survey Sampling International (SSI), which was contracted for this purpose. An earlier but very similar version of the survey experiment was also deployed to students at Arizona State University. In the interest of space I include only the wording from the version deployed to the larger national sample. Replication materials as well as the images used in the instrument are located on the author's website (https://reedmwood.com/).

The experiment varied two factors: gender (all male versus mixed) and relations with civilians (cooperative versus abusive). The text for all conditions follows.

INTRODUCTORY TEXT (SEEN BY ALL GROUPS)

Internal conflicts between states and separatist rebel groups are on the rise globally and are increasingly destabilizing to global security. We will describe one rebel group and ask you a number of questions about how you think the U.S. should react to it. For scientific validity some aspects of the story are fictional. We are asking you to imagine how you would feel about these events if they were happening in the real world today. *Please read the description carefully*, as we will ask you questions about specific details regarding the group.

OPENING TEXT (SEEN BY ALL GROUPS)

For several months, the Kaftarian National Front (KNF), a separatist movement located in South Asia, has been engaged in a violent struggle against the government that has claimed hundreds of lives on both sides. The conflict has rapidly escalated in recent months, and experts believe it has the potential to spill over into neighboring countries, potentially destabilizing the region and creating severe humanitarian costs.

Human rights organizations have expressed serious concerns over the deteriorating human rights situation in the country. They accuse the government of committing many serious human rights abuses and using harsh counterinsurgency tactics that violate international humanitarian law, including the execution of civilians suspected of supporting the KNF and the indiscriminate bombing of villages in KNF-held areas.

TEXT FOR TREATMENT AND CONTROL CONDITIONS (RANDOMLY ASSIGNED)

1. *All male/cooperative*

The KNF, numbering several thousand men, seeks to establish an independent ethnic homeland in which it can practice its cultural and religious traditions without the constant threat of government repression.

International observers have noted that the group generally enjoys positive relations with local civilians: In addition to providing protection for civilians, the KNF has built schools and run healthcare clinics in the areas they control.

In a recent interview for the BBC, Amit, a KNF fighter in his early twenties, asserted that he has dreamed of an independent state since he was child, and he is willing to fight the government, other rebel groups, or anyone else to achieve that dream.

2. *Mixed/cooperative*

The KNF, numbering several thousand men and women, seeks to establish an independent ethnic homeland in which it can practice its cultural and religious traditions without the constant threat of government repression.

International observers have noted that the group generally enjoys positive relations with local civilians: In addition to providing protection for civilians, the KNF has built schools and run healthcare clinics.

More than a third of KNF fighters are female. In a recent interview for the BBC, Amita, a KNF fighter in her early twenties, asserted that she has dreamed of an independent state since she was child, and she is willing to fight the government, other rebel groups, or anyone else to achieve that dream.

3. *All male/abusive*

The KNF, numbering several thousand men, seeks to establish an independent ethnic homeland in which it can practice its cultural and religious traditions without the constant threat of government repression.

International observers have expressed concern over increasing reports of abuse against local civilians by the KNF, including looting of private property and forced recruitment of fighters from local populations.

In a recent interview for the BBC, Amit, a KNF fighter in his early twenties, asserted that he has dreamed of an independent state since he was child, and he is willing to fight the government, other rebel groups, or anyone else to achieve that dream.

4. *Mixed/abusive*

The KNF, numbering several thousand men and women, seeks to establish an independent ethnic homeland in which it can practice its cultural and religious traditions without the constant threat of government repression.

International observers have expressed concern over increasing reports of abuse against local civilians by the KNF, including looting of private property and forced recruitment of fighters from local populations.

More than a third of KNF fighters are female. In a recent interview for the BBC, Amita, a KNF fighter in her early twenties, asserted that she has dreamed of an independent state since she was child, and she is willing to fight the government, other rebel groups, or anyone else to achieve that dream.

DESCRIPTIVE STATISTICS FOR SAMPLES

In this section I present descriptive statistics for the samples used in the analyses presented in chapter 5. Tables C.1 and C.2 present basic summary statistics for the student sample and national sample, respectively. As these statistics demonstrate, while neither sample is completely representative of the U.S. population, there is substantial heterogeneity among the respondents within each sample. This is particularly true of the national sample, though it is worth noting that even in this sample the respondents are substantially more educated than the U.S. population as a whole. Tables C.3 and C.4 present the balance statistics for these samples. Overall, these statistics suggest that reasonable balance in the characteristics of the respondents was achieved across the control and treatment conditions.

Table C.1 Student Sample Summary Statistics

Variable	Mean	Std. Dev.	Min.	Max.
Age				
Under 21	0.295	0.457	0	1
21–30	0.623	0.485	0	1
31–40	0.053	0.225	0	1
41–50	0.016	0.127	0	1
51–60	0.009	0.096	0	1
61 and over	0.002	0.048	0	1
Female	0.408	0.492	0	1
Race (white)	0.594	0.492	0	1
Liberal	0.500	0.501	0	1

Variable	Mean	Std. Dev.	Min.	Max.
Income				
< 25,000	0.224	0.417	0	0
25,000–49,999	0.191	0.394	0	0
50,000–74,999	0.161	0.368	0	0
75,000 and above	0.424	0.495	0	0

N=429

Table C.2 National Sample Summary Statistics

Variable	Mean	Std. Dev.	Min.	Max.
Age				
Under 21	0.032	0.175	0	1
21–30	0.192	0.393	0	1
31–40	0.219	0.414	0	1
41–50	0.146	0.353	0	1
51–60	0.167	0.373	0	1
61–70	0.178	0.382	0	1
71–80	0.059	0.235	0	1
81 and over	0.008	0.089	0	1
Female	0.565	0.496	0	1

(continued)

Table C.2 National Sample Summary Statistics (*continued*)

Variable	Mean	Std. Dev.	Min.	Max.
Race (white)	0.723	0.448	0	1
Education				
Did not complete high school	0.014	0.119	0	1
High school / GED	0.176	0.381	0	1
Some college	0.304	0.460	0	1
Bachelor's degree	0.283	0.450	0	1
Some post-graduate	0.049	0.216	0	1
Graduate degree	0.175	0.379	0	1
Liberal	0.346	0.476	0	1
Income				
< 25,000	0.168	0.374	0	1
25,000–49,999	0.265	0.441	0	1
50,000–74,999	0.215	0.411	0	1
75,000 and above	0.353	0.478	0	1

N=1741

Table C.3 Student Sample Balance Statistics

Variable	Control mean	Treatment mean	Difference in means
Age	1.797 (0.042)	1.859 (0.055)	−0.062
Female	0.373 (0.033)	0.443 (0.034)	−0.070
Race (white)	0.608 (0.033)	0.580 (0.034)	0.028
Liberal	0.493 (0.034)	0.507 (0.034)	−0.014
Income	2.788 (0.081)	2.783 (0.085)	−0.005

Table C.4 National Sample Balance Statistics

Variable	Treatment mean	Control mean	Difference in means
Age	4.057 (0.056)	4.008 (0.057)	0.049
Female	0.549 (0.169)	0.579 (0.167)	−0.031
Race (white)	0.728 (0.015)	0.492 (0.015)	0.010
Education	3.764 (0.046)	3.633 (0.045)	0.131
Liberal	0.351 (0.016)	0.341 (0.016)	0.010
Income	3.764 (0.046)	3.633 (0.045)	0.131

Notes

Introduction

The second epigraph, a statement made by a *miliciana* named Manuela during the Spanish Civil War, is quoted in Lines (2012, 90).

1. Despite relying on samples with different spatial and temporal domains, these datasets reach strikingly similar conclusions about the proportion of armed groups that include female combatants.
2. On this topic see Alison (2009); Baaz and Stern (2009); Coulter (2009); Parashar (2011).
3. This term is commonly employed in the literature on postconflict disarmament, demobilization, and reintegration (DDR). See UN Women (2012, 22–23) for descriptions and discussions of the different categories of women in armed groups.
4. Few studies have attempted to classify women's roles or systematically assess which roles they are likely to play and under what conditions. However, previous studies demonstrate that women play a diverse range of roles within rebel movements (e.g., Lyons 2004; MacKenzie 2009; Viterna 2013). One recent systematic study identifies more than forty specific roles women commonly play in violent revolutionary movements, which the study authors organize into four broad thematic categories: active (e.g., fighting, logistics), support (e.g., smuggling, intelligence), ideological (e.g., propaganda, fundraiser), and caring (e.g., nurse, camp wife) (Vogel, Porter and Kebbell 2014).
5. This definition is very similar to the one used by the Uppsala Conflict Data Program (UCDP) (2018).

6. Conflict scholars have used various death and casualty thresholds to define *civil war* or *civil conflict*. The convention in much of the academic literature on conflict is to refer to conflicts producing more than one thousand deaths per year as *civil wars* and those falling under this threshold as *civil conflicts*. I discuss the specific definitions that help define the sample of cases used in the analyses in chapter 4.
7. Compare, for example, the number of books and articles on female combatants to the number on the onset, duration, or outcome of civil wars; effects of foreign intervention on civil wars; patterns of violence in civil wars; or even rebel fragmentation and alliance formation in civil wars.
8. Examples of these studies include the LTTE (Alison 2009; Herath 2012), ZANLA, and the Zimbabwe People's Revolutionary Army (ZIPRA) (Lyons 2004; O'Gorman 2011), the Sandinistas (Kampwirth 2002; Mason 1992; Reif 1986), Euskadi Ta Askatasuna (ETA) (Hamilton 2007), the RUF (Coulter 2009; MacKenzie 2012), and others.
9. Nonetheless, the prevalence of female combatants in Islamist rebel movements is comparatively low, and those that do participate directly in violence are most likely to serve as suicide bombers (Wood and Thomas 2017). I return to this discussion in chapter 1.
10. Some previous studies have also identified sexual violence as a salient motivator for women's participation in violent extremist groups (Bloom 2011; Speckhard 2008). The explicitly gendered nature of such violence may therefore represent a specific cause of women's recruitment.
11. Mats Utas (2005) refers to this as *tactical* agency. By contrast, *strategic* agency requires the ability to forecast future conditions and to make use of others' tactical agency.
12. Indeed, men are comparatively more likely to be targeted for coercion and violence during war (Carpenter 2005).
13. Viterna notes that members of the movement considered many of the support roles women occupied (particularly in logistics, communications, and medicine) as more prestigious than frontline combat positions. Thus, while a clear gender bias existed, it is not necessarily the case that women in the FMLN were systematically excluded from more desirable roles.
14. Similar divisions are also present in many social movements (Cable 1992).
15. Interestingly, many rebel groups appear to allow marriage but require approval from commanders. The requirement that commanders, who are mostly male, grant approval for marriage raises additional concerns regarding gender bias, asymmetric relationships, and the possibility of forced marriages.
16. Notable exceptions include Queen Boudica (60–61 CE) in Roman Britain, the female warriors of the Dahomey Kingdom of West Africa (ca. 1650–1850), and female soldiers in the USSR (1941–1945). See J. Goldstein (2001).

17. According to Laura Sjoberg (2014, 40), thirty countries permit women to serve in combat; however, she provides no source for this number. Even if this higher estimate is accurate, it represents only about 15 percent of national armed forces globally.
18. Though see Walters (2015) for a critique of this study.
19. Despite the widespread acceptance that gender is a continuum, the narrative assumes that it is binary and largely overlaps with biological sex.
20. Historical, psychological, and sociological evidence supports some of these claims. For instance, existing studies suggest that men who are perceived as more "masculine" are seen as more attractive and possess distinct advantages in the selection of romantic partners (Puts, Jones, and DeBruine 2012). Moreover, this relationship may be strongest in times of conflict or crisis (DeBruine et al. 2010; Brooks et al. 2011).
21. Not all women supported the actions of the White Feather Brigades; moreover, the evidence suggests that both the organization of the brigades and much of the gendered propaganda of World War I was the creation of male politicians and military officers (see Gullace 1997).

1. Why Rebels Mobilize Women for War

1. Despite privileging demand-side forces, I nonetheless acknowledge the role supply-side factors can play in this process. All else equal, an increase in the supply of women willing to fight for a group should exert a positive influence on their prevalence in armed groups because any expansion in the supply of potential recruits reduces the effort rebel leaders spend locating viable recruits and reduces the marginal value of the incentives rebels must offer to attract new recruits.
2. Previous studies likewise suggest that high levels of coercive or indiscriminate violence against civilians may ultimately prove counterproductive (e.g., Fortna 2015).
3. A similar expansion of the age range for eligible draftees occurred in the United States during World War II.
4. Under certain conditions rebel leaders may also view child soldiers as preferable to adults (Beber and Blattman 2013). Notably, child soldiering has been roundly condemned by the international community in recent years. The practice has been considered a war crime under international law since the adoption of the Rome Statute (Article 8(2)(b)(xxvi)). Moreover, in 2012 the UN Security Council approved a resolution (11 in favor; 0 opposed; 4 abstentions) calling for naming and shaming of groups that utilize child soldiers and recommending sanctions on those groups.

5. These strategies are not necessarily mutually exclusive.
6. As noted in the introduction, at least some portion of the female population in most countries possesses the physical, mental, and emotional qualities necessary to competently participate in combat.
7. The movement also launched a campaign to increase male fighters' and civilians' acceptance of female combatants in its ranks (Chris Johnson 1992, 160).
8. For example, Kriger (1992, 196–198) describes the contentious relationship between ZANLA guerrillas and local chiefs and headmen. Moreover, she asserts that the rebels frequently exploited peasant hostilities toward these local elites as a strategy to gain control over villages. Moreover, Kesby (1996) discusses the tensions between the guerrillas and local elders. In particular, government counterinsurgency strategies, dislocation, and rising violence disempowered traditional patriarchs, providing rebels with greater authority over the population.
9. This pattern also appeared to reflect geographic variations in conflict intensity: the National Liberation Front (NLF) was more likely to train, arm, and deploy women in combat in areas where the fighting was most intense.
10. While exact figures are unknown, many thousands of women served in combat (acting as snipers, combat pilots, etc.) in both gender-segregated and integrated combat units.
11. The discussion linking group ideology to female recruitment draws heavily on arguments advanced in previously published co-authored work with Jakana Thomas (Wood and Thomas 2017).
12. The role of leaderships' political ideology in civil conflict has frequently been ignored or discounted in much of the literature examining and explaining rebel group strategy, tactics, and organization. However, recent studies have sought to highlight the potential importance of political ideology in explaining both group-level strategy formation and individual behaviors during armed conflicts (e.g., Gutiérrez Sanín and Wood 2014, 2017; Schubiger and Zelina 2017; Ugarriza and Craig 2012).
13. This does not imply that women are inherently subordinate or subjugated in all premodern or non-Western cultures. However, traditional gender-based divisions of labor—in which women are expected to primarily fulfill maternal or domestic obligations—have historically resulted in the exclusion of women from power and the denial of equal rights and opportunities in society. On the relationship between traditional gender roles, development, and gender equality, see Inglehart and Norris (2003, chap. 2).
14. See Wood and Thomas (2017) for a more focused discussion and empirical analysis of the relationship between Islamist ideologies and female combatants.
15. Of course, implementation of these laws proved challenging in many areas, and the party was not always willing to enforce them.

16. While this goal often includes (directly or indirectly) efforts to reshape gender hierarchies, Marxist leaders' true commitment to egalitarianism often fails to match their rhetoric (see Lobao 1990).
17. Data compiled by Jessica Maves Braithwaite and Kathleen Gallagher Cunningham (2018) demonstrate that the majority of rebel groups that originate from these sorts of organizations espouse some form of leftist or communist ideology.
18. While the leaders of these movements heavily emphasized Marxist rhetoric regarding national liberation, they were, for the most part, neither particularly knowledgeable about nor deeply committed to Marxist ideology (see Dobell 2000, 57–60 [SWAPO]; Kriger 1992, 95–96 [ZANLA]; Sullivan 1988, 269 [ETA]).

2. The Strategic Implications of Female Fighters

1. Only a handful of studies have considered the potential impact of female combatants on rebel behavior or their influence on any aspect of the conflicts in which they fight. Among these very few examples, Lindsey O'Rourke (2009) finds that suicide bombings carried out by females tend to produce greater numbers of causalities than bombings conducted by their male counterparts. Meredith Loken (2017) finds that the presence of female combatants fails to lessen the incidence of wartime rape committed by armed groups.
2. Some studies find that rebel military capabilities exert a nonlinear effect on the odds of negotiated settlement (e.g., Hultquist 2013). This somewhat counterintuitive observation obtains because rebel demands are endogenous to their capabilities and their expectations of victory: as the group survives longer and gains more power, it feels emboldened to demand greater concessions from the state (Bapat 2005). Regardless, more powerful rebels are less likely to suffer defeat and more likely to eventually achieve some of their goals.
3. While the strength of gender stereotypes and perceptions of sex- and gender-based differences vary cross-culturally, remarkable similarities nonetheless exist across diverse social and cultural contexts (see Costa, Terracciano, and McRae 2001; Williams and Best 1990).
4. Sex stereotypes also result from observing the roles men and women are most likely to assume in society. Thus, the relationship between stereotypes and roles are at least partly endogenous (Eagly and Steffen 1984).
5. While assumptions of women's roles in war and how they are depicted may vary across societies, there is substantial evidence of cross-cultural similarity. For example, although images of women in war are culturally distinct,

Lorraine Bayard de Volo (2001, 37) found strong similarities between dominant themes in developed Western countries and those in Latin America.

6. This does not suggest that rebel leaders that recruit female combatants have become gender egalitarian or that they reject previously held gender biases. However, it likely implies that the strength of these attitudes have weakened relative to the leadership's preference for meeting its recruitment demands.
7. GAM relied heavily on strategies oriented toward attracting international attention to their cause as its leadership recognized the low likelihood of victory without significant international support for the group or without international pressure on the Indonesian government to pursue a negotiated settlement to the conflict (Aspinall 2009, 227–228).
8. Lisa Margaret Lines (2012, 158) reports that while only some 2 percent of the Republican combat forces were women, photos with female fighters constituted as much as 50 percent of all photos related to the war in some issues of *Ahora*, a left-wing newspaper affiliated with the organization Juventudes Socialists Unificadas (JSU) in the 1930s.
9. The perceived novelty of women's participation in armed groups may increase the more the roles they perform diverge from those traditionally sanctioned by their societies. Jelena Batinic (2015, 125–126), for example, asserts that while female nurses in the Yugoslav partisan forces were visible because of their sheer numbers, female combatants were conspicuous because they were a novelty.
10. These films are also available on other video-hosting sites as well as the producers' websites, likely making the total viewership much higher than reported here.
11. These include clips from ABC News, CNN, and PBS *Newshour.*
12. Coverage of the Kurdish groups is often in English; the majority of videos available online related to FARC are not.
13. This characterization also extends to other Kurdish factions, such as PJAK in Iran and Iraq and the Peshmerga forces of northern Iraq.
14. The authors also note that British media tended to highlight the appearance and sexuality of female fighters. This is common in news reports of women participating in violence (see Lavie-Dinur, Karniel, and Azran 2015; Melzer 2009).
15. By contrast, Sjoberg (2018) argues that while stories focused on PKK/YPG women accentuate their agency and celebrate their roles, those stories focused on "jihadi brides" and other female supporters of ISIL remove the agency of these women and serve to delegitimize and vilify the group.
16. Interestingly, public sentiment and military policy turned against female fighters after 1937, particularly as the Republican forces attempted to transform

from a guerrilla force to a more disciplined conventional force (see Nash 1993; Lines 2009).

17. The politics of labeling is important in armed conflicts (see Barrinha 2011; Bhatia 2008). Governments typically attempt to characterize rebels as brigands, bandits, or terrorists; conversely, rebels attempt to frame themselves as legitimate freedom fighters. The labels applied to the group need not reflect the true goals or actual behaviors of the group. Rather, they are part of each side's efforts to construct a narrative that they use to shape the beliefs and preferences of specific audiences.
18. This implies that those groups that exclude women entirely are unlikely to become aware of their potential benefits. Moreover, given the arguments outlined in chapter 1, groups that exclude female combatants due to fears of backlash from their domestic constituencies would still be unlikely to receive such benefits. For example, because the audiences on which fundamentalist religious rebel movements rely are more likely to view the inclusion of women in the armed group as inappropriate or anathema to their beliefs, they would also be unlikely to view female combatants as sympathetic figures or to see the groups that utilize them as legitimate.
19. Few studies have explicitly investigated the use of gender in rebel propaganda, let alone done so in a rigorous, systematic manner. One exception, however, is Loken's (2018) analysis, which focuses largely on the use of gendered imagery in the propaganda produced and disseminated by the PIRA.
20. SWAPO produced a number of publications during its war for independence. These included *Namibia Today*, *SWAPO Information Bulletin*, *Namibia News*, *Namibian Woman*, and *The Combatant*.
21. Numerous additional examples of images of female combatants used in rebel propaganda material or materials produced by movements allied with a rebellion can be found in Miescher and Henrichsen (2004) and on the websites of the African Activist Archive (http://africanactivist.msu.edu/index.php), the Palestine Poster Project Archives (https://www.palestineposterproject.org), OSPAAAL (http://www.ospaaal.com), the Dogma Collection (http://dogmacollection.com), Docs Populi (http://www.docspopuli.org/Archives.html), and the Oakland Museum of California (http://collections.museumca.org/).
22. I would like to extend special thanks to Lincoln Cushing of Docs Populi for sharing many of these images and for providing invaluable information on their origins and context.
23. This sympathy largely eroded after the infamous Belsan attack and hostage crisis (Speckhard 2008, 1000).
24. The underlying constraints on the supply of female potential fighters are closely related to those that lead rebel leaders to resist women's inclusion in

combat. Principally, prevailing gender norms, biases, and patterns of socialization shape the behaviors, attitudes, and preferences of both men and women. Much as these factors compel men to support and participate in violence, they often lead women to avoid it, thus reducing their willingness to become fighters in armed resistance movements. Moreover, while some women possess the physical capabilities to endure the rigors of combat, many more do not. Thus, the total number of women willing and able to fight is typically lower than the number of men.

25. It is also important to note that the deployment of women in noncombat roles can also serve to increase the number of men available for combat. Large military organizations require a substantial number of members to serve in noncombat roles such as logistics, transportation, training, recruitment, and so on. Allowing women to serve in these roles frees up men to serve in combat roles. Thus, even deploying women in noncombat roles should substantially increase the group's military capabilities. Nonetheless, allowing women to serve in both roles provides an additional number of combatants above what could normally be produced in an all-male force.
26. Relatedly, Reyko Huang (2016a) reports that 39 percent of rebel groups in her sample of civil wars engaged in some form of explicit diplomatic efforts with foreign states, indicating the frequency with which rebel leaders attempt to engage the international community and solicit external support.
27. While it is possible to critique the EPLF's actual commitment to gender equality and women's rights, these issues were central to their platform during the conflict (Matsuoka and Sorenson 2001, 122).
28. Numerous other organizations, both secular and religious, mobilized in opposition to U.S. funding for the war in El Salvador (see Nepstad 2004). While not directly supporting the FMLN, these movements tended to view the FMLN as a product of state repression and often indirectly advocated on their behalf. At the minimum, it is clear that they viewed U.S. support to the regime as morally indefensible and lobbied to end it.

3. Female Combatants in Three Civil Wars

1. This includes the YPG in Syria and Partiya Jiyana Azad a Kurdistanê (PJAK) in Iran. The PKK officially maintains that these groups constitute separate but "closely linked" affiliates. However, some Kurdish fighters have publically asserted that all of the groups are members of the broader PKK organization (Bradley and Parkinson 2015). Due to their close linkages, the discussions in this chapter broadly refer to each of these groups as well as their

subsidiary women's units within these groups, such as the Yekîneyên Jinên Azad ên Star (YJA-STAR) and Yekîneyên Parastina Jin (YPJ).

2. For information on women's status, rights, and opportunities in each of the cases, see Turkey/Kurds (van Bruinessen 2001; Yüceşahin and Özgür 2008; Yüksel 2006, 783); Sri Lanka/Tamils (Fernando 1974; Herath 2012, 50–55; Jayatilaka and Amrithalingam 2015; Langford 1981); Zimbabwe (Jacobs 1983; Kriger 1992, 75–78; Wells 2003). Data on fertility rates, labor force participation, and access to secondary education from the World Bank (2018) in the years leading up to the conflict also reflect systematic marginalization in these cases.
3. Sri Lanka represents a partial exception. Sri Lankan women generally appeared to enjoy a comparatively high level of equality with men in the years prior to the conflict. World Bank (2018) indicators suggest that Sri Lanka's male-female education ratio, female literacy rate, and fertility rate were better than the global average in the years preceding the conflict. It is therefore possible that this factor at least partly explains the eventual high participation of women in the LTTE. Yet, as with the other cases, the integration of women in to the group's combat force occurred slowly over time.
4. Cansiz remained a member of the organization until her murder in Paris in 2013.
5. See, for example, Bengio (2016), as well as the organizations' websites.
6. By roughly 1995 the group formally organized an armed women's wing that eventually became known as the Free Women Units (YJA-STAR) (see Celebi 2010; White 2015, 140–142).
7. In almost all cases, it is difficult to trace the change in recruitment strategy to a specific year. In each of the case illustrations I identify a window during which it is clear that the leadership began to recruit female combatants, deploy women in combat roles, or dramatically increase the number of female combatants.
8. These estimates reflect the total battle deaths in the conflict and as such include government, rebel, and civilian fatalities.
9. Values for the year 1993 are not available in the Global Terrorism Database. For more information, see "Frequently Asked Questions," Global Terrorism Database, http://www.start.umd.edu/gtd/faq/.
10. This reflects "one-sided violence," which the UCDP defines as direct and intentional lethal violence against noncombatants.
11. This search was conducted on March 8, 2018. The search included the terms "women," "girl," and "female" in addition to the group names "YPG," "YPJ," and "PKK." I also removed all moderately similar stories from the results. I limited the search to headlines because these stories would be most visible to

readers. Removing the headline restriction increases the number of articles returned to nearly one thousand.

12. As additional evidence of the salience (if not popularity) of Kurdish female fighters in the West, in 2014 clothing producer H&M briefly sold an outfit that to many appeared to be inspired by the uniforms of the YPG. After a social media backlash, the seller apologized and claimed that the similarity was coincidental (see Bulos 2014).
13. What is perhaps most remarkable about this specific case is that much of the narrative is fictional. The woman from the famous Rehana photo is real, and she did serve in an auxiliary home guard unit affiliated with the YPJ. However, there is no indication she ever participated in combat, and many other elements of the narrative appear to be fictional or highly embellished (Varghese 2015). As such, the "Rehana" persona seems to be an amalgamation of many different female combatants in the YPJ and PKK.
14. See, for example, PKK (https://www.pkkonline.net/en/); YPG (https://www.ypgrojava.org/english); and PJAK (http://www.pajk-online.com/).
15. "Meet Delila, Fighting in the Rajo Front of the Resistance," People's Defense Units (YPG) website, March 5, 2018, https://www.ypgrojava.org/Meet-Delila%2C-fighting-in-Rajo-front-of-the-resistance; "AFP Video: Syria Kurd Women Fighters Out for Revenge Against Jihadists," People's Defense Units (YPG) website, November 11, 2016, https://www.ypgrojava.org/AFP-Video%3A-Syria-Kurd-women-fighters-out-for-revenge-against-jihadists-; "Who are the Women's Protection Units?," People's Defense Units (YPG) website, September 26, 2016, https://www.ypgrojava.org/Who-are-the-Womens-Protection-Units.
16. See "About the Kurdish Project," The Kurdish Project, https://thekurdish-project.org/about-the-kurdish-project/.
17. See Females on the Frontline, http://femalesonthefrontline.org/.
18. See Phillips and Berkell (2016) and "Remove the PKK from the Government's List of Proscribed Terrorist Organizations," Petition to UK Parliament, closed May 26, 2016, https://petition.parliament.uk/archived/petitions/113738.
19. The proportion of female combatants among the Black Tigers is reportedly somewhat higher than their proportion in the group's combat forces more generally (Stack-O'Connor 2007a, 53).
20. The death counts included in the figure include all actors in the conflict, which in this case also includes deaths attributed to and accruing to the other militant Tamil factions such as the EPRLF and the Tamil Eelam Liberation Organization (TELO) as well as the LTTE.
21. It is also around this time that the LTTE began relying on child soldiers (Herath 2012, 77), which, as discussed in chapter 1, is also closely related to the demand for troops.

22. See Balasingham (1993).
23. According to Arnestad, the Tamil leadership agreed to her conditions that they would not censor or prescreen the film for approval. Given the tight control the leadership maintained over access to the group and information about it, the assertion about the absence of preconditions is either inaccurate, or the LTTE were confident that a portrayal of its female members would necessarily present the group in a positive light. The latter would be consistent with the observation that the media never (or only rarely) provided negative coverage of female combatants in the group. See Arnestad (2009).
24. There is no evidence that the rate of women's participation declined as the conflict went on; if anything, it is likely that women's participation increased through the 1990s.
25. The UCDP External Support Dataset (Högbladh, Pettersson, and Themnér 2011) lists 1991 as the first year in which the LTTE received substantial financial resources from the diaspora community.
26. The intended audience could also have been the domestic population, in which case the intention would likely have been to shame men living in Sri Lanka to join the fight. It is less likely that the intention was simply to motivate other women to fight, given that the letter writer expressly highlights the cowardice of the male recipient of the letter. Of course, the letter may not have been directed to a single audience, and, regardless of the target, the intention of juxtaposing the bravery of the female author and the cowardice of the émigré male is clear.
27. Of course, some men may decide to return to home to fight. The assumption here is that the barriers to do so are substantially greater than the barriers for men already present in the conflict state.
28. Mike Kesby (1996) notes that the majority of women involved in the armed struggle were civilian activists, not combatants.
29. This number is based on a combination of numeric estimates and estimated proportions. For example, Norma Kriger (1992, 191) suggests that ZANLA had some two thousand trained female combatants, which would represent about 10 percent of the total force based on the estimated twenty thousand total troops in 1979. Similarly, Tanya Lyons (2004, 59) asserts that roughly 10 percent of ZIPRA's fighting force was composed of female combatants.
30. There is less information overall on women's participation in ZIPRA.
31. The decision to arm and train female guerillas resulted partly from women's claims that their assignments to combat support roles already frequently exposed them to the risk of armed encounters with enemy forces (Lyons 2004, 109–110; O'Gorman 2011, 58).
32. It is worth noting that there was an apparent divide between the younger leadership and the "old guard" (e.g., Mugabe) in that the former were more

committed to socialism, while the latter were more focused on nationalism (Kriger 2002, 92–93). Ultimately, the latter group prevailed in the power struggle among the factions.

33. Some critics have accused the group of blatant hypocrisy, pointing out that while ZANU officially espoused an ideology of gender equality in practice, it subjugated and exploited women within the organization (Nhongo-Simbanegavi 2000, 1–4, 135–137).
34. It should be noted that Josephine Nhongo-Simbanegavi is skeptical of the extent to which women were ever employed in "actual combat."
35. Though, readers may wish to see Mazarire (2017) on ZANU's efforts to secure and increase diplomatic support from socialist states, particularly in the final years of the conflict.

4. Empirical Evaluation of Female Combatant Prevalence

1. Nor do I attempt to explain or evaluate women's motives for participation in rebellion. I view investigations into women's motives for joining armed groups as being complementary to explaining rebel leaders' decisions to recruit women. For analyses that focus more directly on the strategies that women adopt during war, how they negotiate their relationships with armed actors, and their experiences within rebel movements, I suggest ethnographic studies such as Akawa (2014), Coulter (2009), Hamilton (2007), Herath (2012), Kampwirth (2002), Lyons (2004), MacKenzie (2012), Nash (1995), O'Gorman (2011), Taylor (1999), Utas (2005), and Viterna (2013).
2. All materials necessary to replicate the following analyses are available on the author's website (https://reedmwood.com/).
3. Updates to the original Wood and Thomas (2017) dataset are discussed in appendix A.
4. WARD also includes a brief narrative and source list for each case. Samples of the coding narratives are included in appendix B. Complete narratives are available on the project website: https://reedmwood.com/home-page/women-in-armed-rebellion-dataset-ward/.
5. These include the Fuerzas Populares de Liberacíon Farabundo Martí (FPL), the Ejército Revolcionario del Pueblo (ERP), Resistencia Nacional (RC), Partido Comunista Salvadoreño (PCS), and Partido Revolucionario de los Trabajadores Centroamericanos (PRTC).
6. In most of these cases, not only did coders fail to locate any specific references to women's participation in the group, they were generally unable to locate any descriptive information about the group other than a rough sketch of its goals, areas of operation, and reports of battles with government forces.

7. The United Nations now advocates for the expansion of the definition of *combatant* to include women who served in support roles in rebel groups within DDR programs because women included in the latter category are often denied the benefits associated with DDR simply because they did not carry weapons.
8. When estimated female troop numbers are reported, the proportion of the group they represent is estimated by dividing that number by the total number of combatants reported in the Nonstate Actors (NSA) Dataset (Cunningham, Gleditsch, and Salehyan 2009, 2013).
9. Sixty cases receive different scores on these measures, representing 23 percent of the sample. However, the polychoric correlation between the "best" and "high" measures of the variable is 0.963.
10. In all cases the results are similar in both statistical significance and substantive effect.
11. I combine the cases that terminated between 1964 and 1969 with those terminating in the subsequent decade because of the relatively small number of observations.
12. The Frente Popular de Liberación de Saguía el Hamra y Río de Oro (POLISARIO) in Western Sahara, Morocco represents a potential exception to the general absence of female fighters in North Africa. Many women served in the movement's military wing as nurses, administrators, and political leaders. There is little evidence, however, that female members of POLISARIO participated in combat, though some women apparently received military training (see Zunes and Mundy 2010, 133).
13. Information on active conflict years is taken from the UCDP Dyadic Dataset.
14. Specifically, I back coded the ideology indicators to 1964 using the same coding criteria. I also revised the coding for a handful of the existing cases based on new information.
15. For instance, rightist ideologies are not included as a distinct category because only two included in the sample were coded this way, which impedes statistical analysis. Groups that embraced right-wing or anti-communist ideologies, such as União Nacional para a Independência Total de Angola (UNITA) and the Contras, are instead folded into the category *secular (other).*
16. Specifically, I use the *start date* variable included in the UCDP Dyadic Dataset, which records (as accurately as possible) the year of the first battle-related death. Conflicts are coded as *terminated* when the violence falls below twenty-five annual deaths.
17. Dara Kay Cohen's dataset (2013a) uses a different sample of civil conflict cases than the one used for WARD. I therefore matched the conflict cases in Cohen's dataset to the groups included in WARD using the conflict descriptions Cohen provides. For cases included in this sample but absent from

Cohen's dataset, I coded whether or not the rebel groups involved in the conflict ever employed forced recruitment or abductions using the same criteria and sources applied by Cohen.

18. Thomas and Bond (2015) included a group-specific forced recruitment variable, but it is only available for rebel groups located in Africa.
19. As a robustness check, I also employed the average values of the indicator over the duration of the conflict. In alternative models I also substitute the ratio of female-male secondary school enrollment as a measure of gender equality. The results are substantively similar in both cases.
20. A Brant test reveals that some of the covariates violate the proportional odds assumption. As a robustness check, I therefore reran all models using ordinary least squares (OLS) estimation. The results of these models are highly consistent with those presented below.
21. I conducted separate analyses for each interaction.
22. It is important to note that interactions are inherently symmetrical and should be examined as such (Berry, Golder, and Milton 2012). In this case, it is necessary to examine both the conditional influence of nationalism on other ideologies as well as the conditional influence of other ideologies on nationalism. The results provide support for the latter relationship but not the former.
23. However, there is some evidence that rebels may choose to recruit children explicitly because they present specific strategic benefits. Principally, children are more easily manipulated and controlled (Beber and Blattman 2013).
24. In the sample used herein, 72 percent of the groups who utilized suicide bombings espoused an Islamist ideology. More than half of the groups within that subset employed women to carry out at least some of these attacks.

5. Empirical Evaluation of the Effects of Female Combatants

1. An alternative indicator using the value of the *democracy* variable at the outset of the conflict produces very similar results.
2. For additional information on the external support variables, readers should consult Salehyan, Gleditsch, and Cunningham (2011).
3. While the data do not identify the specific organizations supporting the rebel groups, the codebook associated with the dataset provides a general description of the constituencies. Many of the constituencies identified in the codebook are global religious movements and ethno-national diaspora communities (e.g., Irish Americans, Eritreans, or Kurds). I make one modification to this measure. Numerous sources (e.g., Högbladh, Pettersson, and Themnér 2011; Matsuoka and Sorenson 2001) indicate that the EPLF received substantial

support from the Eritrean diaspora as well as from other external groups. I therefore code this group as having a transnational constituency.

4. I rely on the results using the measure of *female combatant prevalence* that excludes cases in which women served only as suicide bombers because this measure is more consistent with the arguments presented in chapter 2. However, as the results indicate, the relationship is similar regardless of the specific measure.
5. A cursory examination of the factors predicting the use of forced recruitment supports this claim. Regression results suggest that variables such as *conflict severity* and *duration* are among the most predictive of the use of forced recruitment by the rebels.
6. Based on the change in logged scale of the dependent variable.
7. The marginal effects for the prevalence variable are very similar for groups that engage in forced recruitment as well as those that do not. Moreover, there is no evidence that this variable conditions the relationship between female combatants and rebel troop size.
8. Dr. Manekin is currently an assistant professor at the Hebrew University of Jerusalem. Additional analyses and results from this experiment and related experiments appear elsewhere in co-authored work with Dr. Manekin. See, for example, Manekin and Wood 2019.
9. Survey Sampling International (SSI) was contracted to deploy the survey to the national sample. The ASU sample was gathered with the assistance of the ASU School of Politics and Global Studies Experimental Lab.
10. The specific wording of the instrument is included in appendix C.
11. Hence, the experiment had a 2 × 2 factorial design.
12. While the dependent variable reflects discrete ordered categories, I rely on OLS for ease of interpretation of the effect of the treatment, as is common in experimental analyses using Likert scale indicators as the dependent variable. Nonetheless, the results of alternative models using ordered probit estimation are highly similar in terms of substantive effect and significance.
13. This indicator is recoded from a multicategory indicator ranging from "extremely liberal" to "extremely conservative." Using the complete scale does not change the results for the other indicators; however, it masks the statistically significant relationship between self-reported liberal ideology and support for the rebel movement.
14. The experiment deployed to the national sample included a larger number of income categories. However, in order to maintain consistency across the samples, I use the same four categories included in the earlier survey deployed to the student sample. The results are virtually identical when using the full range of income categories.

15. The results presented here exclude respondents that completed the survey in less than 48 percent of the median time for the sample. I exclude these respondents, which constitute roughly 5 percent of the total number of original respondents in the sample, because they were unlikely to have actually read the material or considered their responses given the speed with which they completed the assignment. The results are very similar, though somewhat weaker, when including these respondents.
16. However, there are numerous studies that systematically assess the correlates of state support for foreign rebel groups (e.g., Salehyan, Gleditsch, and Cunningham 2011; Salehyan 2009).
17. This sample includes only eight cases of rebel organizations that received tangible support from a diaspora community, which almost certainly reflects underreporting. One way to look at this is that these cases are perhaps the most overt examples of diaspora funding and may therefore have the largest impact. Regardless, these results should be treated as preliminary and exploratory rather than representative.
18. Results of these models produce highly similar results.
19. As discussed in the previous chapters, while *religious* and *leftist* are mutually exclusive categories, the category *nationalist* is inclusive. Thus, a group can be religious and nationalist or leftist and nationalist. This explains why the percentages listed above do not add up to 100 percent.
20. Replacing *nationalist* with the previously employed measure *secular non-Marxist*, which produces mutually exclusive ideology categories, produces similar results. The only notable difference is that *leftist* is no longer statistically significant.

References

Achvarina, Vera. 2010. "Child Soldiering in Intrastate Conflicts: An Empirical Analysis." Ph.D. dissertation. University of Pittsburgh. http://d-scholarship.pitt.edu/8034/1/ETD_Achvarina.pdf.

Achvarina, Vera, and Simon Reich. 2006. "No Place to Hide: Refugees, Displaced Persons, and the Recruitment of Child Soldiers." *International Security* 31 (1): 127–164.

Adamson, Fiona. 2013. "Mechanisms of Diaspora Mobilization and the Transnationalization of Civil War." In *Transnational Dynamics of Civil War*, ed. Jeffrey Checkel, 63–88. Cambridge: Cambridge University Press.

Akawa, Matha. 2014. *The Gender Politics of the Namibian Liberation Struggle*. Basel, Switzerland: Basler Afrika Bibliographien.

Alexander, David, and Phil Stewart. 2015. "US Military Opens All Combat Roles to Women." *Reuters*, December 3, 2015. http://www.reuters.com/article/us-usa-military-women-combat-idUSKBN0TM28520151203.

Alexander, Jocelyn, and JoAnn McGregor. 2004. "War Stories: Guerrilla Narratives of Zimbabwe's Liberation War." *History Workshop Journal* 57 (1): 79–100.

Alison, Miranda. 2003. "Cogs in the Wheel? Women in the Liberation Tigers of Tamil Eelam." *Civil Wars* 6 (4): 37–54.

——. 2004. "Women as Agents of Political Violence: Gendering Security." *Security Dialogue* 35 (4): 447–463.

——. 2009. *Women and Political Violence: Female Combatants in Ethno-national Conflict*. New York: Routledge.

Allansson, Marie, Erik Melander, and Lotta Themner. 2016. "Organized Violence, 1989–2016." *Journal of Peace Research* 54 (4): 574–587.

Anagnostopoulou, Margaret Poulos. 2001. "From Heroines to Hyenas: Women Partisans During the Greek Civil War." *Contemporary European History* 10 (3): 481–501.

Andvig, Jens Christopher, and Scott Gates. 2010. "Recruiting Children for Armed Conflict." In *Child Soldiers in the Age of Fractured States*, ed. Scott Gates and Simon Reich, 77–92. Pittsburgh: University of Pittsburgh Press.

Annan, Jeannie, Christopher Blattman, Dyan Mazurana, and Kristopher Carlson. 2009. "Women and Girls at War: 'Wives,' Mothers, and Fighters in the Lord's Resistance Army." Unpublished manuscript. https://pdfs.semanticscholar.org/9cce/c59c0422fe953e70039371a298ffa36813cc.pdf.

Arnestad, Beate. 2009. "My Daughter, the Terrorist." *Amanpour* (blog), CNN, December 14. http://amanpour.blogs.cnn.com/2009/12/14/my-daughter-the-terrorist/.

Asal, Victor H., and R. Karl Rethemeyer. 2015. Big Allied and Dangerous Dataset (BAAD), v. 2. www.start.umd.edu/baad/database.

Aspinall, Edward. 2007. "From Islamism to Nationalism in Aceh, Indonesia." *Nations and Nationalism* 13 (2): 245–263.

——. 2009. *Islam and Nation: Separatist Rebellion in Aceh, Indonesia.* Singapore: NUS Press.

Baaz, Maria Eriksson, and Maria Stern. 2009. "Why Do Soldiers Rape? Masculinity, Violence, and Sexuality in the Armed Forces in the Congo (DRC)." *International Studies Quarterly* 53 (2): 495–518.

Balasingham, Adele Ann. 1993. *Women Fighters of Liberation Tigers.* Jaffna, Sri Lanka: Thasan Printers. http://tamilnation.co/books/Eelam/adeleann.htm.

Balch, Lindsay, Andrew Enterline, and Kyle Joyce. 2008. "Third-Party Intervention and the Civil War Process." *Journal of Peace Research* 45 (3): 345–363.

Bapat, Navin. 2005. "Insurgency and the Opening of the Peace Process." *Journal of Peace* Research 42 (6): 699–171.

Barnes, Tiffany, and Emily Beaulieu. 2014. "Gender Stereotypes and Corruption: How Candidates Affect Perceptions of Election Fraud." *Politics and Gender* 10 (3): 365–391.

Barrinha, André. 2011. "The Political Importance of Labeling: Terrorism and Turkey's Discourse on the PKK." *Critical Studies on Terrorism* 4 (2): 163–180.

Barter, Shane. 2014. *Civilian Strategy in War: Insights from Indonesia, Thailand, and the Philippines.* New York: Palgrave.

——. 2015. "Zones of Control and Civilian Strategy in the Aceh Conflict." *Civil Wars* 17 (3): 340–356.

Batinic, Jelena. 2015. *Women and Yugoslav Partisans: A History of World War II Resistance.* Stanford, CA: Stanford University Press.

Bayard de Volo, Lorraine. 2001. *Mothers of Heroes and Martyrs: Gender Identity Politics in Nicaragua, 1979–1999*. Baltimore, MD: Johns Hopkins University Press.

BBC. 2013. "Colombia FARC Rebels Launch Website of the Female Rebel." October 13, 2013. http://www.bbc.com/news/world-latin-america-24510335.

——. 2014a. "#BBCtrending: Who Is the 'Angel of Kobane'?" November 3, 2014. http://www.bbc.com/news/blogs-trending-29853513.

——. 2014b. "Islamic State Are Afraid to See Women with Guns." September 5, 2014. https://www.youtube.com/watch?v=TGVbpsGmLVo.

——. 2014c. "Who Are the Female Fighters of the PKK?" January 5, 2014. http://www.bbc.com/news/world-middle-east-25610424.

——. 2016. "Women to Serve in Close Combat Roles in British Military." July 8, 2016. http://www.bbc.com/news/uk-36746917.

Bear, Stephen, Noushi Rahman, and Corinne Post. 2010. "The Impact of Board Diversity and Gender Composition on Corporate Responsibility and Firm Reputation." *Journal of Business Ethics* 97 (2): 207–221.

Beber, Bernd, and Christopher Blattman. 2013. "The Logic of Child Soldiering and Coercion." *International Organization* 67: 65–104.

Beit-Hallahmi, Benjamin. 2005. "The Return of Martyrdom: Honour, Death and Immortality." In *Religious Fundamentalism and Political Extremism*, ed. Leonard Weinberg and Ami Pedahzur, 11–34. London: Frank Cass.

Ben-David, Yuval. 2017. "Israel's Women Combat Soldiers on the Frontline of Battle for Equality." *Reuters*, May 7, 2017. http://www.reuters.com/article/us-israel-women-idUSKBN16E23P.

Ben Shitrit, Lihi. 2016. *Righteous Transgression: Women's Activism on the Israeli and Palestinian Right*. Princeton, NJ: Princeton University Press.

Bengio, Ofra. 2016. "Game Changers: Kurdish Women in Peace and War." *Middle East Journal* 1 (Winter): 30–46.

Bernal, Victoria. 2000. "Equality to Die For? Women Guerrilla Fighters and Eritrea's Cultural Revolution." *Political and Legal Anthropology Review* 23: 61–76.

Berry, William, Matt Golder, and Daniel Milton. 2012. "Improving Tests of Theories Positing Interactions." *Journal of Politics* 74 (3): 653–671.

Bhatia, Michael. 2008. "Fighting Words: Naming Terrorists, Bandits, Rebels, and Other Violent Actors." *Third World Quarterly* 26 (1): 5–22.

Bishopsgate Institute. 2018. "Committee for Freedom in Mozambique, Angola and Guine (CFMAG)." http://www.bishopsgate.org.uk/Library/Special-Collections-and-Archives/Protest-and-Campaigning/Committee-for-Freedom-in-Mozambique-Angola-and-Guine-CFMAG.

Blaydes, Lisa, and Drew A. Linzer. 2008. "The Political Economy of Women's Support for Fundamentalist Islam." *World Politics* 60 (4): 576–609.

Blee, Kathleen. 1996. "Becoming a Racist: Women in Contemporary Ku Klux Klan and Neo-Nazi Groups." *Gender and Society* 10 (6): 680–702.

Bloom, Mia. 2011. *Bombshells: Women and Terrorism*. Philadelphia: University of Pennsylvania Press.

Bob, Clifford. 2005. *The Marketing of Rebellion: Insurgents, Media, and International Activism*. Cambridge: Cambridge University Press.

Bouta, Tsjeard, Georg Ferks, and Ian Bannon. 2005. *Gender Conflict and Development*. Washington, DC: World Bank.

Bradley, Matt, and Joe Parkinson. 2015. "America's Marxist Allies Against ISIS." *Wall Street Journal*, July 24, 2015. https://www.wsj.com/articles/americas-marxist-allies-against-isis-1437747949.

Bradley-Engen, Mindy, Keely Damphousse, and Brent Smith. 2009. "Punishing Terrorists: A Re-examination of US Federal Sentencing in the Postguidelines Era." *International Criminal Justice Review* 19 (4): 433–455.

Braithwaite, Jessica Maves, and Kathleen Gallagher Cunningham. 2018. "When Organizations Rebel: Introducing the Foundations of Rebel Group Emergence (FORGE) Dataset." Unpublished manuscript.

Brooks, Robert, Isabel Scott, Alexei Makalov, Michael Kasumovie, Andrew Clark, and Ian Penton-Voak. 2011. "National Income Inequality Predicts Women's Preference for Masculinized Faces Better than Health Does." *Proceedings of the Royal Society B: Biological Sciences* 278: 810–812.

Brown, Heather. 2012. *Marx on Gender and the Family: A Critical Study*. Leiden, Netherlands: Brill.

Brun, Cathrine. 2005. "Women in the Local/Global Fields of War and Displacement in Sri Lanka." *Gender, Technology and Development* 9 (1): 57–80.

Bulos, Nabih. 2014. "Radical Chic? Kurds Say H&M Jumpsuit Mimics Fighter Garb." *Los Angeles Times*, October 8, 2014. https://www.latimes.com/world/middleeast/la-fg-hm-jumpsuits-kurdish-fighter-garb-20141007-story.html.

Bumiller, Elisabeth, and Thom Shanker. 2013. "Pentagon Is Set to Lift Combat Ban for Women." *New York Times*, January 23, 2013. http://www.nytimes.com/2013/01/24/us/pentagon-says-it-is-lifting-ban-on-women-in-combat.html.

Byman, Daniel, Peter Chalk, Bruce Hoffman, William Rosenau, and David Brannan. 2001. *Trends in Outside Support for Insurgent Movements*. Santa Monica, CA: Rand Corporation.

Cable, Sherry. 1992. "Women's Social Movement Involvement: The Role of Structural Availability in Recruitment and Participation Processes." *Sociological Quarterly* 33 (1): 35–50.

Campbell, D'Ann. 1993. "Women in Combat: The World War II Experience in the United States, Great Britain, Germany, and the Soviet Union." *Journal of Military History* 57 (2): 301–323.
Caprioli, Mary. 2005. "Primed for Violence: The Role of Gender Inequality in Predicting Internal Conflict." *International Studies Quarterly* 49 (2): 161–178.
Card, Kathryn. 2011. "Comparative Study of Chechen and Kurdish Female Terrorists in the Mass Media." Master's thesis, George Mason University. http://digilib.gmu.edu/jspui/bitstream/handle/1920/10761/Card_thesis_2016.pdf?sequence=1&isAllowed=y.
Carpenter, R. Charli. 2005. "'Women, Children, and Other Vulnerable Groups': Gender, Strategic Frames, and the Protection of Civilians as a Transnational Issue." *International Studies Quarterly* 49: 295–334.
Carter, Brenda, Kevan Insko, David Loeb, and Marlene Tobias, eds. 1989. *A Dream Compels Us: Voices of Salvadoran Women*. Boston: South End Press.
Casey, Nicholas. 2016. "In a Rebel Camp in Colombia, Marx and Free Love Reign." *New York Times*, March 18, 2016. http://www.nytimes.com/2016/03/19/world/americas/colombia-farc-rebels.html?hp&action=click&pgtype=Homepage&clickSource=image&module=photo-spot-region®ion=top-news&WT.nav=top-news#.
Cawkill, Paul, Alison Rogers, Sarah Knight, and Laura Spear. 2009. *Women in Ground Close Combat Roles: The Experience of Other Nations and a Review of Academic Literature*. Fareham, UK: Defense Science and Technology Laboratory. DSTL/CR37770 V3–0.
Celebi, Erdogan. 2010. "Female Separatism: The Role of Women in the PKK/KONGRA-GEL Terrorist Organization." In *Terrorism and the Internet: Threats, Target Groups, Deradicalisation Strategies*, ed. Hans-Liudger Dienel, Yair Sharan, Christian Rapp, and Niv Ahituv, 105–114. Berlin: IOS Press.
Cheibub, José Antonio, Jennifer Gandhi, and James Raymond Vreeland. 2010. "Democracy and Dictatorship Revisited." *Public Choice* 143 (2): 67–101.
Chicago Project on Security and Terrorism. 2015. Suicide Attack Database. Data file. September 30. http://cpostdata.uchicago.edu/.
Chinchilla, Norma Stoltz. 1983. "Women in Revolutionary Movements: The Case of Nicaragua." In *Revolution in Central America*, ed. Stanford Central American Action Network, 422–434. Boulder, CO: Westview Press.
——. 1990. "Revolutionary Popular Feminism in Nicaragua: Articulating Class, Gender, and National Sovereignty." *Gender and Society* 4 (3): 370–397.
Chuchryk, Patricia. 1991. "Women in the Revolution." In *Revolution and Counterrevolution in Nicaragua*, ed. Thomas Walker, 143–165. Boulder, CO: Westview Press.
Chung, Fay. 2006. *Reliving the Second Chimurenga: Memories from the Liberation Struggle in Zimbabwe*. Stockholm: The Nordic Africa Institute.

——. 2016. "The Role of Female Fighters in the Liberation War." *Financial Gazette*, September 15, 2016. http://www.financialgazette.co.zw/role-of-female-fighters-in-the-liberation-war/.

Clapham, Christopher. 1995. "The International Politics of African Guerrilla Movements." *South African Journal of International Affairs* 3 (1): 81–91.

Clayton, Govinda. 2013. "Relative Rebel Strength and the Onset and Outcome of Civil War Mediation." *Journal of Peace Research* 50 (5): 609–622.

Cock, Jacklyn. 1991. *Colonels and Cadres: War and Gender in South Africa*. Cape Town, South Africa: Oxford University Press.

Coggins, Bridgette. 2011. "Friends in High Places: International Politics and the Emergence of States from Secessionism." *International Organization* 65 (3): 433–467.

——. 2015. "Rebel Diplomacy: Theorizing Violence Non-state Actors' Strategic Use of Talk." In *Rebel Governance in Civil War,* ed. Ana Arjona, Nelson Kasfir, and Zachariah Mampilly, 98–118. Cambridge: Cambridge University Press.

Cohen, Dara Kay. 2013a. "Explaining Rape During Civil War: Cross-national Evidence (1980–2009)." *American Political Science Review* 107 (3): 461–477.

——. 2013b. "Female Combatants and the Perpetration of Violence: Wartime Rape in the Sierra Leone Civil War." *World Politics* 65 (3): 383–415.

——. 2016. *Rape During Civil War*. Ithaca, NY: Cornell University Press.

Collier, Paul, and Anke Hoeffler. 2004. "Greed and Grievance in Civil War." *Oxford Economic Papers* 56: 563–595.

Cook, David. 2005. "Women Fighting in Jihad." *Studies in Conflict and Terrorism* 28 (5): 375–384.

Cordero, Isabel Coral. 1998. "Women in War: Impact and Response." In *Shining and Other Paths: War and Society in Peru, 1980–1995*, ed. Steve J. Stern, 345–374. Durham, NC: Duke University Press.

Costa, Paul, Antonio Terracciano, and Robert McRae. 2001. "Gender Differences in Personality Traits Across Cultures: Robust and Surprising Findings." *Journal of Personality and Social Psychology* 81 (2): 322–331.

Coulter, Chris. 2009. *Bush Wives and Girl Soldiers: Women's Lives Through War and Peace in Sierra Leone*. Ithaca, NY: Cornell University Press.

Coulter, Chris, Mariam Persson, and Mats Utas. 2008. *Young Female Fighters in African Wars*. Uppsala, Sweden: Nordiska Afrikainstitutet.

Cragin, Kim, and Sara Daly. 2009. *Women as Terrorists: Mothers, Recruiters and Martyrs*. Santa Barbara, CA: Praeger Security International.

Criss, Nur Bilge. 1996. "The Nature of PKK Terrorism in Turkey." *Studies in Conflict and Terrorism* 18 (1): 17–37.

Cunningham, David, Kristian Gleditsch, and Idean Salehyan. 2009. "It Takes Two: A Dyadic Analysis of Civil War Duration and Outcome." *Journal of Conflict Resolution* 53 (3): 570–597.

——. 2013. "Non-state Actors in Civil Wars: A New Dataset." *Conflict Management and Peace Science* 30 (5): 516–531.

Cunningham, Karla. 2003. "Cross-regional Trends in Female Terrorism." *Studies in Conflict and Terrorism* 26 (3): 171–195.

——. 2009. "Female Survival Calculations in Politically Motivated Settings: How Political Violence and Terrorism as Viewed as Pathways to Life." *Studies in Conflict and Terrorism* 32 (7): 561–575.

Dahal, Swechchha. 2015. "Challenging Boundaries: The Narratives of Female Ex-Combatants in Nepal." In *Female Combatants in Conflict and Peace: Challenging Gender in Violence and Post-Conflict Reintegration*, ed. Seema Shekhawat, 185–199. New York: Palgrave MacMillan.

Daly, Kathleen, and Rebecca Bordt. 1995. "Sex Effects and Sentencing: An Analysis of the Statistical Literature." *Justice Quarterly* 12 (1): 141–175.

Davies, Sara, and Jacqui True. 2015. "Reframing Conflict-related Sexual and Gender-based Violence: Bringing Gender Analysis Back In." *Security Dialogue* 46 (6): 495–512.

Davis, Jessica. 2013. "Evolution of the Global Jihad: Female Suicide Bombers in Iraq." *Studies in Conflict and Terrorism* 36 (4): 279–291.

Davis, Shannon, and Theodore Greenstein. 2009. "Gender Ideology: Components, Predictors, and Consequences." *Annual Review of Sociology* 35: 87–105.

DeBruine, Lisa, Benedict Jones, John Crawford, Lisa Welling, and Anthony Little. 2010. "The Health of a Nation Predicts Their Mate Preferences: Cross-cultural Variation in Women's Preferences for Masculinized Male Faces." *Proceedings of the Royal Society B: Biological Sciences* 277: 2405–2410.

De Grazia, Victoria. 1992. *How Fascism Ruled Italian Women: Italy, 1922–1945*. Berkeley: University of California Press.

Del Re, Emanuela C. 2015. "Female Combatants in the Syrian Conflict, in the Fight Against or with ISIS, and in the Peace Process." In *Female Combatants in Conflict and Peace: Challenging Gender in Violence and Post-Conflict Reintegration*, ed. Seema Shekhawat, 84–99. New York: Palgrave MacMillan.

Denov, Myriam, and Christine Gervais. 2007. "Negotiating (In)Security: Agency, Resistance, and Resourcefulness among Girls Formerly Associated with Sierra Leone's Revolutionary United Front." *Signs* 32 (4): 885–910.

De Pauw, Linda Grant. 1998. *Battle Cries and Lullabies: Women in War from Prehistory to the Present*. Norman: University of Oklahoma Press.

DeVotta, Neil. 2009. "The Liberation Tigers of Tamil Eelam and the Lost Quest for Separatism in Sri Lanka." *Asian Survey* 49 (6): 1021–1051.

Dirik, Dilar. 2014. "Western Fascination with 'Badass' Kurdish Women." *Al-Jazeera*, October 29, 2014. http://www.aljazeera.com/indepth/opinion/2014/10/western-fascination-with-badas-20141021124105277736.html.

Dobell, Lauren. 2000. *SWAPO's Struggle for Namibia, 1960–1991: War by Other Means*. 2nd ed. Basel, Switzerland: P. Schlettwein.

Donnell, John. 1967. "Vietcong Recruitment: Why and How Men Join." Memorandum: RM-5486–1-ISA/ARPA. Santa Monica, CA: Rand Corporation.

Drake, C. J. M. 1998. "The Role of Ideology in Terrorists' Target Selection." *Terrorism and Political Violence* 10 (2): 53–85.

Durham, Martin. 1998. *Women and Fascism*. London: Routledge.

El-Bushra, Judy. 2003. "Fused in Combat: Gender Relations and Armed Conflict." *Development in Practice* 13 (2–3): 252–265.

Eagly, Alice, and Valerie Steffen. 1984. "Gender Stereotypes Stem from the Distribution of Men and Women into Social Roles." *Journal of Personality and Social Psychology* 46 (4): 735–754.

Eagly, Alice, and Wendy Wood. 2011. "Social Role Theory." *Handbook of Theories of Social Psychology* 2: 458–478.

Eck, Kristine. 2009. "From Armed Conflict to War: Ethnic Mobilization and Conflict Intensification." *International Studies Quarterly* 53 (2): 369–388.

——. 2014. "Coercion in Rebel Recruitment." *Security Studies* 23 (2): 364–398.

Eggert, Jennifer Philippa. 2018. "Female Fighters During the Lebanese Civil War: Individual Profiles, Pathways, and Motivations." *Studies in Conflict and Terrorism*. https://doi.org/10.1080/1057610X.2018.1529353.

Elshtain, Jean B. 1987. *Women and War*. Chicago: Chicago University Press.

Emerson, Michael, and David Hartman. 2006. "The Rise of Religious Fundamentalism." *Annual Review of Sociology* 32: 127–144.

Enloe, Cynthia. 1983. *Does Khaki Become You?* London: South End Press.

——. 1999. *Bananas, Beaches and Bases: Making Feminist Sense of International Politics*. Berkeley: University of California Press.

Epstein, Cynthia Fuchs. 1989. *Deceptive Distinctions: Sex, Gender and the Social Order*. New Haven, CT: Yale University Press.

Epstein, Yoram, Ran Yanovich, Daniel Moran, and Yuval Heled. 2013. "Physiological Employment Standards IV: Integration of Women in Combat Units." *European Journal of Applied Physiology* 113: 267–269.

Fearon, James. 1994. "Rationalist Explanations for War." *International Organization* 49 (3): 379–414.

Fernando, Dallas. 1974. "A Note on Differential Fertility in Sri Lanka." *Demography* 11 (3): 441–456.

Finestone, Aharon, Charles Milgrom, Ran Yanovich, Rachel Evans, Naama Constantini, and Daniel Moran. 2014. "Evaluation of the Performance of Women as Light Infantry Soldiers." *Biomed Research International*. http://dx.doi.org/10.1155/2014/572953.

Fisher, Max. 2013. "Map: Which Countries Allow Women in Frontline Combat Roles?" Worldviews, *Washington Post*, January 25, 2013. https://www.wash

ingtonpost.com/news/worldviews/wp/2013/01/25/map-which-countries-allow-women-in-front-line-combat-roles/?utm_term=.2766a120f18d.

Fortna, Page. 2015. "Do Terrorists Win? Rebels' Use of Terrorism and Civil War Outcomes." *International Organization* 69 (3): 519–556.

France24. 2013. "Slain PKK Member Was a Rebel with a Cause." http://www.france24.com/en/20130111-kurdish-france-slain-pkk-female-turkey-sakine-cansiz-profile-legend.

Freeden, Michael. 1998. "Is Nationalism a Distinct Ideology?" *Political Studies* 46 (4): 748–765.

Gates, Scott. 2002. "Recruitment and Allegiance: The Microfoundations of Rebellion." *Journal of Conflict Resolution* 46 (1): 111–130.

——. 2017. "Membership Matters: Coerced Recruitment and Rebel Allegiance." *Journal of Peace Research* 54 (5): 674–686.

Gautam, Shobha, Amrita Banskota, and Rita Manchanda. 2001. "Where There Are No Men: Women in the Maoist Insurgency in Nepal." In *Women, War, and Peace in South Asia*, ed. Rita Manchanda, 214–251. New Delhi: Sage.

Geisler, Gisela. 1995. "Troubled Sisterhood: Women and Politics in Southern Africa: Case Studies from Zambia, Zimbabwe, and Botswana." *African Affairs* 94 (377): 545–578.

——. 2004. *Women and the Remaking of Politics in Southern Africa: Negotiating Autonomy, Incorporation, and Representation*. Uppsala, Sweden: Nordiska Afrikainstitutet.

Gellner, Ernest. 2008. *Nations and Nationalism*. 2nd ed. Ithaca, NY: Cornell University Press.

Gent, Stephen. 2008. "Going in When It Counts: Military Intervention and the Outcome of Civil Conflicts." *International Studies Quarterly* 53 (4): 713–735.

Gentzkow, Matthew, and Jessie M. Shapiro. 2010. "What Drives Media Slant? Evidence from US Daily Newspapers." *Econometrica* 78 (1): 35–71.

Gleditsch, Kristian Skrede. 2002. "Expanded Trade and GDP Data." *Journal of Conflict Resolution* 46 (5): 712–724.

Goldstein, Joshua. 2001. *War and Gender: How Gender Shapes the War System and Vice Versa*. Cambridge: Cambridge University Press.

Goldstein, Leslie. 1980. "Mill, Marx and Women's Liberation." *Journal of the History of Philosophy* 18 (3): 319–334.

Gonzalez-Perez, Margaret. 2006. "Guerrilleras in Latin America: Domestic and International Roles." *Journal of Peace Research* 43 (3) 313–329.

——. 2008. *Women and Terrorism: Female Activity in Domestic and International Terror Groups*. New York: Routledge.

——. 2011. "The False Islamization of Female Suicide Bombers." *Gender Issues* 28: 50–65.

Goswami, Namrata. 2015. *Indian National Security and Counter-Insurgency: The Use of Force vs. Non-violent Response*. New York: Routledge.

Grasmick, Harold, Linda Wilcox, and Sharon Bird. 1990. "The Effects of Religious Fundamentalism and Religiosity on Preference for Traditional Family Norms." *Sociological Inquiry* 60 (4): 352–369.

Gullace, Nicoletta. 1997. "White Feathers and Wounded Men: Female Patriotism and the Memory of the Great War." *Journal of British Studies* 36 (2): 178–206.

Gunes, Cengiz. 2012. *The Kurdish National Movement in Turkey: From Protest to Resistance.* London: Routledge.

Gurr, Ted Robert. 1993. "Why Minorities Rebel: A Global Analysis of Communal Mobilization and Conflict Since 1945." *International Political Science Review* 14 (2): 161–202.

Gutiérrez Sanín, Francisco, and Francy Carranza Franco. 2017. "Organizing Women for Combat: The Experience of the FARC in the Colombian War." *Journal of Agrarian Change* 17: 770–778.

Gutiérrez Sanín, Francisco, and Elisabeth Jean Wood. 2014. "Ideology in Civil War: Instrumental Adoption and Beyond." *Journal of Peace Research* 51 (2): 213–226.

——. 2017. "What Should We Mean by 'Pattern of Political Violence'? Repertoire, Targeting, Frequency, and Technique." *Perspectives on Politics* 15 (1): 20–41.

Hall, John. 2014. "Rehana Is Alive and Well." *Daily Mail*, October 28, 2014. http://www.dailymail.co.uk/news/article-2810780/Rehana-alive-ISIS-fanatics-NOT-beheaded-Poster-girl-Kurdish-freedom-fighters-escaped-Kobane-hellhole-friends-tell-MailOnline.html.

Hamilton, Carrie. 2007. *Women and ETA: The Gender Politics of Radical Basque Nationalism*. Manchester, UK: Manchester University Press.

Harbom, Lotta, Erik Melander, and Peter Wallensteen. 2008. "Dyadic Dimensions of Armed Conflict, 1946–2007." *Journal of Peace Research* 45 (5): 697–710.

Heger, Lindsay, and Danielle Jung. 2015. "Negotiating with Rebels: The Effect of Rebel Service Provision on Conflict Negotiations." *Journal of Conflict Resolution* 61 (6): 1203–1229 .

Hegghammer, Thomas. 2013. "The Recruiter's Dilemma: Signaling and Rebel Recruitment Tactics." *Journal of Peace Research* 50 (1): 3–16.

Hellmann-Rajanayagam, Dagmar. 1994. *Tamil Tigers: Armed Struggle for Identity.* Stuttgart, Germany: Franz Steiner Verlag.

——. 2008. "Female Warriors, Martyrs and Suicide Attackers: Women in the LTTE." *International Review of Modern Sociology* 34 (1): 1–25.

Henshaw, Alexis Leanna. 2016a. "Where Women Rebel: Patterns of Women's Participation in Armed Rebel Groups: 1990–2008." *International Feminist Journal of Politics* 18 (1): 39–60.

——. 2016b. "Why Women Rebel: Greed, Grievance, and Women in Armed Rebel Groups." *Journal of Global Security Studies* 1 (3): 204–219.

——. 2017. *Why Women Rebel: Understanding Women's Participation in Armed Rebel Groups.* New York: Routledge.

Herath, Tamara. 2012. *Women in Terrorism: Case of the LTTE.* New Delhi: Sage.

Herrera, Natalia, and Douglas Porch. 2008. "'Like Going to a Fiesta'—The Role of Female Fighters in Colombia's FARC-EP." *Small Wars and Insurgencies* 19 (4): 609–634.

Hertel, Bradley, and Michael Hughes. 1987. "Religious Affiliation, Attendance, and Support for 'Pro-Family' Issues in the United States." *Social Forces* 65: 858–882.

Högbladh, Stina, Therésa Pettersson, and Lotta Themnér. 2011. "External Support in Armed Conflict: 1975–2009." UCDP External Support Dataset, v. 1.0. http://ucdp.uu.se/downloads/extsup/ucdp_external_support_primary_warring_party_codebook_1.0.pdf.

Hoover Green, Amelia. 2016. "The Commander's Dilemma: Creating and Controlling Armed Group Violence." *Journal of Peace Research* 53 (5): 619–632.

Hopgood, Stephen. 2005. "Tamil Tigers, 1987–2002." In *Making Sense of Suicide Missions*, ed. Diego Gambetta, 43–76. Cambridge: Cambridge University Press.

Horne, Christine, Pazit Ben-Nun Bloom, Kyle Irwin, Dan Miodownik, and Michael Hechter. 2016. "The Legitimacy of Alien Rulers." *Swiss Political Science Review* 22 (4): 454–469.

Huang, Reyko. 2016a. "Rebel Diplomacy in Civil War." *International Security* 40 (4): 89–126.

——. 2016b. *The Wartime Origins of Democratization: Civil War, Rebel Governance, and Political Regimes.* Cambridge: Cambridge University Press.

Hultman, Lisa. 2007. "Battle Losses and Rebel Violence: Raising the Costs for Fighting." *Terrorism and Political Violence* 19 (2): 205–222.

Hultquist, Philip. 2013. "Power Parity and Peace? The Role of Relative Power in Civil War Settlement." *Journal of Peace Research* 50 (5): 623–634.

Humphreys, Macartan, and Jeremy Weinstein. 2008. "Who Fights? The Determinants of Participation in Civil War." *American Journal of Political Science* 52 (2): 436–455.

Inglehart, Ronald, and Pippa Norris. 2003. *Rising Tide: Gender Equality and Cultural Change.* Cambridge: Cambridge University Press.

Israelsen, Shelli. 2018. "Why Now? Timing Rebel Recruitment of Female Combatants." *Studies in Conflict and Terrorism.* https://doi.org/10.1080/1057610X.2018.1445500.

Jacobs, Susie. 1983. "Women and Land Resettlement in Zimbabwe." *Review of African Political Economy* 10 (27/28): 33–50.

Jacques, Karen, and Paul Taylor. 2009. "Female Terrorism: A Review." *Terrorism and Political Violence* 21 (3): 499–515.

Jancar, Barbara. 1981. "Women in the Yugoslav National Liberation Movement: An Overview." *Studies in Comparative Communism* 14 (2): 144–164.

Jaulola, Marjaana. 2013. *Post-tsunami Reconstruction in Indonesia: Negotiating Normativity through Gender Mainstreaming Initiative in Aceh.* New York: Routledge.

Jayatilaka, Danesh, and Kopalapillai Amirthalingam. 2015. *The Impact of Displacement on Dowries in Sri Lanka.* Washington, DC: Brookings Institution.

Jo, Hyeran. 2015. *Compliant Rebels: Rebel Groups and International Law in World Politics.* Cambridge: Cambridge University Press.

Johnson, Cedric. 2003. "From Popular Anti-imperialism to Sectarianism: The African Liberation Support Committee and Black Power Radicals." *New Political Science* 25 (4): 477–507.

Johnson, Chris. 1992. *Women on the Frontline: Voices from Southern Africa.* London: MacMillan Academic.

Johnson, Kay Ann. 1983. *Women, the Family and Peasant Revolution in China.* Chicago: Chicago University Press.

Johnson, Loch. 2006. *Strategic Intelligence: Understanding the Hidden Side of Government.* Westport, CT: Greenwood.

Jones, Benjamin, and Elenora Mattiacci. 2017. "A Manifesto in 140 Characters or Fewer: Social Media as a Tool of Rebel Diplomacy." *British Journal of Political Science.* https://doi.org/10.1017/S0007123416000612.

Jones, David, and Susanne Stein. 2008. "Southern Africa: Twentieth Century." In *The Oxford Encyclopedia of Women in World History, Volume 1*, ed. Bonnie Smith. Oxford: Oxford University Press. Online edition.

Jordan, Kim, and Myriam Denov. 2007. "Birds of Freedom? Perspectives on Female Emancipation and Sri Lanka's Liberation Tigers of Tamil Elam." *Journal of International Women's Studies* 9 (1) 42–62.

Jost, John, Christopher Frederico, and Jamie Napier. 2009. "Political Ideology: Its Structure, Functions, and Elective Affinities." *American Review of Psychology* 60: 307–337.

Kafanov, Lucy. 2016. "How All-Female ISIS Morality Police 'Khansa Brigade' Terrorize Mosul." *NBC News*, November 20, 2016. http://www.nbcnews.com/storyline/isis-uncovered/how-all-female-isis-morality-police-khansaa-brigade-terrorized-mosul-n685926.

Kaiser, Daniel. 2017. "'Makers of Bonds and Ties': Transnational Socialisation and the National Liberation of Mozambique." *Journal of Southern Africa Studies* 43 (1): 29–48.

Kalyvas, Stathis. 2008. "Ethnic Defection in Civil War." *Comparative Political Studies* 41 (8): 1043–1068.

Kalyvas, Stathis, and Matthew Kocher. 2007. "How 'Free' Is Free Riding in Civil Wars? Violence, Insurgency, and the Collective Action Problem." *World Politics* 59 (2): 177–216.

Kampwirth, Karen. 2001. "Women in the Armed Struggle in Nicaragua: Sandinistas and Contras Compared." In *Radical Women in Latin America: Left and Right*, ed. Victoria Gonzalez and Karen Kampwirth, 79–101. University Park: Pennsylvania State University Press.

——. 2002. *Women and Guerrilla Movements: Nicaragua, El Salvador, Chiapas, Cuba*. University Park: Pennsylvania State University Press.

——. 2004. *Feminism and the Legacy of Revolution: Nicaragua, El Salvador, Chiapas*. Athens: Ohio University Press.

Karim, Sabrina. 2017. "Restoring Confidence in Post-conflict Security Sectors: Survey Evidence from Liberia on Female Ratio Balancing Reforms." *British Journal of Political Science*. https://doi.org/10.1017/S0007123417000035.

Keating, Joshua. 2012. "Where a Woman's Place Is on the Frontlines." *Foreign Policy* 10 (February). http://foreignpolicy.com/2012/02/10/where-a-womans-place-is-on-the-front-lines/.

Kesby, Mike. 1996. "Arenas of Control, Terrains of Gender Contestation: Guerrilla Struggle and Counter Insurgency Warfare in Zimbabwe, 1972–1980." *Journal of Southern African Studies* 22 (4): 561–584.

Kesic, Obrad. 1999. "Women and Gender Imagery in Bosnia: Amazons, Sluts, Victims, Witches, and Wombs." In *Gender Politics in the Western Balkans: Women and Society in Yugoslavia and the Yugoslav Successor States*, ed. Sabrina Ramet and Branka Magas, 187–202. University Park: Pennsylvania State University Press.

Kössler, Reinhart, and Henning Melber. 2002. "The West German Solidarity Movement with the Liberation Struggles of Southern Africa. A (Self-)Critical Retrospective." In *Germany's Africa Policy Revisited*, ed. Ulf Engel and Robert Kappel, 103–126. Münster, Germany: Lit Verlag.

Kreutz, Joakim. 2010. "How and When Armed Conflicts End: Introducing the UCDP Conflict Termination Dataset." *Journal of Peace Research* 47 (2): 243–250.

Kriger, Norma J. 1992. *Zimbabwe's Guerrilla War: Peasant Voices*. New York: Cambridge University Press.

Krystalli, Roxanne. 2016. "Why Love in the FARC Isn't So Free (You Wouldn't Know It From Reading the New York Times)." Monkey Cage, *Washington Post*, March 24, 2016. https://www.washingtonpost.com/news/monkey-cage/wp/2016/03/24/women-in-the-farc-have-had-a-mixed-experience-you-wouldnt-know-that-from-the-new-york-times/.

Lacina, Bethany, and Nils Petter Gleditsch. 2005. "Monitoring Trends in Global Combat: A New Dataset of Battle Deaths." *European Journal of Population* 21 (2–3): 145–166.

Lahoud, Nelly. 2014. "The Neglected Sex: The Jihadis' Exclusion of Women from Jihad." *Terrorism and Political Violence* 26 (5): 780–802.

Lamb, Robert. 2014. "Rethinking Legitimacy and Illegitimacy: A New Approach to Assessing Support and Opposition across Disciplines." A Report of the Center for Strategic and International Studies (CSIS) Program on Crisis, Conflict, and Cooperation. https://www.csis.org/analysis/rethinking-legitimacy-and-illegitimacy.

Lamothe, Dan. 2016. "Army's First Female Infantry Officer Is Capt. Kristen Griest, Ranger School Graduate." *Washington Post*, April 27. https://www.washingtonpost.com/news/checkpoint/wp/2016/04/27/armys-first-female-infantry-officer-is-capt-kristen-griest-ranger-school-graduate/?utm_term=.2c399e7c7cf5.

Langford, C. M. 1981. "Fertility Change in Sri Lanka since the War: An Analysis of the Experience of Different Districts." *Population Studies* 35 (2): 285–306.

Lasley, Trace, and Clayton Thyne. 2015. "Secession, Legitimacy and the Use of Child Soldiers." *Conflict Management and Peace Science* 33 (3): 289–308.

Lavie-Dinur, Avit, Yuval Karniel, and Tal Azran. 2015. "'Bad Girls': The Use of Gendered Media Frames in the Israeli Media's Coverage of Israeli Female Political Criminals." *Journal of Gender Studies* 24 (3): 326–346.

Leites, Nathan, and Charles Wolf. 1970. *Rebellion and Authority: An Analytic Essay on Insurgent Conflicts*. Chicago: Markham.

Lichbach, Mark Irving. 1998. *The Rebel's Dilemma*. Ann Arbor: University of Michigan Press.

Lines, Lisa Margaret. 2009. "Female Combatants in the Spanish Civil War: Milicianas on the Front Lines and in the Rearguard." *Journal of International Women's Studies* 10 (4): 168–187.

——. 2012. *Milicianas: Women in Combat in the Spanish Civil War*. Lanham, MD: Lexington Press.

Lischer, Sarah. 2005. *Dangerous Sanctuaries: Refugee Camps, Civil War, and the Dilemmas of Humanitarian Assistance*. Ithaca, NY: Cornell University Press.

Lobao, Linda. 1990. "Women in Revolutionary Movements: Changing Patterns of Latin American Guerilla Struggle." *Dialectical Anthropology* 15: 211–232.

Loken, Meredith. 2017. "Rethinking Rape: The Role of Women in Wartime Violence." *Security Studies* 26 (1): 60–92.

——. 2018. "Women in War: Militancy, Legitimacy and Rebel Success." Ph.D. dissertation. University of Washington.

Lorber, Judith. 1994. *Paradoxes of Gender*. New Haven, CT: Yale University Press.

Luciak, Ilja. 2001a. *After the Revolution: Gender and Democracy in El Salvador, Nicaragua and Guatemala*. Baltimore: Johns Hopkins University Press.

——. 2001b. "Gender Equality, Democratization, and the Revolutionary Left in Central America: Guatemala in Comparative Perspective." In *Radical Women in Latin America: Left and Right*, ed. Victoria Gonzalez and Karen Kampwirth, 189–211. University Park: Pennsylvania State University Press.

Lyons, Tanya. 2004. *Guns and Guerrilla Girls: Women and the Zimbabwean National Liberation Struggle*. Trenton, NJ: Africa World Press.

Macdonald, Sharon. 1987. "Drawing Lines—Gender, Peace, and War: An Introduction." In *Images of Women in Peace and War: Cross-Cultural and Historical Perspectives*, ed. Sharon Macdonald, Pat Holden, and Shirley Ardener, 1–26. Madison: University of Wisconsin Press.

MacKenzie, Megan. 2009. "Securitization and Desecuritization: Female Soldiers and the Reconstruction of Women in Post-Conflict Sierra Leone." *Security Studies* 18 (2): 241–261.

——. 2012. *Female Soldiers in Sierra Leone: Sex, Security, and Post-Conflict Development*. New York: New York University Press.

Mahmood, Saba. 2005. *Politics of Piety: The Islamic Revival and the Feminist Subject*. Princeton, NJ: Princeton University Press.

Malet, David. 2013. *Foreign Fighters: Transnational Identity in Civil Conflicts*. Oxford: Oxford University Press.

Mampilly, Zachariah. 2011. *Rebel Rulers: Insurgent Governance and Civilians Life During War*. Ithaca, NY: Cornell University Press.

Manchanda, Rita. 2004. "Maoist Insurgency in Nepal: Radicalizing Gendered Narratives." *Cultural Dynamics* 16 (2/3): 237–258.

Manekin, Devorah, and Reed M. Wood. 2019. "The Influence of Female Fighters on External Audience Attitudes Toward Armed Rebellion." Unpublished manuscript.

Marcus, Aliza. 2007. *Blood and Belief: The PKK and the Kurdish Fight for Independence*. New York: New York University Press.

Marshall, Andrew. 2002. "The Widow's Battalion." *New York Times Magazine*, January 20, 2002, 30–31.

Marshall, Monty G., Ted Robert Gurr, and Keith Jaggers. 2011. *Polity IV Project: Political Regime Characteristics and Transitions, 1800–2013, Dataset Users' Manual*. Center for Systemic Peace.

Mason, T. David 1992. "Women's Participation in Central American Revolutions: A Theoretical Perspective." *Comparative Political Studies* 25 (1): 63–89.

Mason, T. David, and Patrick J. Fett. 1996. "How Civil Wars End: A Rational Choice Approach." *Journal of Conflict Resolution* 40 (4): 546–568.

Mason, T. David, and Dale A. Krane. 1989. "The Political Economy of Death Squads: Toward a Theory of the Impact of State-sanctioned Terror." *International Studies Quarterly* 33 (2): 175–198.

Matsuoka, Atsuko, and John Sorenson. 2001. *Ghosts and Shadows: Construction of Identity and Community in an African Diaspora*. Toronto: University of Toronto Press.

Matthews, Michael, Morten Enders, Janice Laurence, and David Rohall. 2009. "Role of Group Affiliation and Gender Attitudes Toward Women in the Military." *Military Psychology* 21: 241–251.

Mayer, Tamar. 2002. "Gender Ironies of Nationalism: Setting the Stage." In *Gender Ironies of Nationalism: Sexing the Nation*, ed. Tamar Mayer, 1–24. New York: Routledge.

Mazarire, Gerald Chikozho. 2017. "ZANU's External Networks 1963–1979: An Appraisal." *Journal of Southern African Studies* 43 (1): 83–106.

Mazurana, Dyan, Susan McKay, Kristopher Carlson, and Janel Kasper. 2002. "Girls in Fighting Forces and Groups: Their Recruitment, Participation, Demobilization, and Reintegration." *Peace and Conflict: Journal of Peace Psychology* 8 (2): 97–123.

McKay, Susan. 2004. "Reconstructing Fragile Lives: Girls' Social Reintegration in Northern Uganda and Sierra Leone." *Gender and Development* 12 (3): 19–30.

——. 2005. "Girls as 'Weapons of Terror' in Northern Uganda and Sierra Leonean Rebel Fighting Forces." *Studies in Conflict and Terrorism* 28: 385–397.

Melzer, Patricia. 2009. "'Death in the Shape of a Young Girl: Feminist Responses to Media Representations of Women Terrorists during the 'German Autumn' of 1977." *International Feminist Journal of Politics* 11 (1): 35–62.

Miescher, Giorgio, and Dag Henrichsen. 2004. *African Posters: A Catalogue of the Poster Collection in the Basler Afrika Bibliographien*. Basel, Switzerland: Basler Afrika Bibliographien.

Miroff, Nick. 2016. "See the Female Fighters Who Strike Fear into the Hearts of Colombian Troops." *Washington Post*, January 31, 2016. https://www.washingtonpost.com/news/worldviews/wp/2016/01/31/see-the-female-fighters-who-strike-fear-in-the-hearts-of-the-colombian-army/.

Moghadam, Assaf. 2008. "Motives for Martyrdom: Al-Qaida, Salafi Jihad, and the Spread of Suicide Attacks." *International Security* 3 (33): 46–78.

Moghadam, Valentine. 2003. *Modernizing Women: Gender and Social Change in the Middle East*. Boulder, CO: Lynne Reiner Press.

Molyneux, Maxine. 1985. "Mobilization with Emancipation? Women's Interests, the State and Revolution in Nicaragua." *Feminist Studies* 11 (2): 227–254.

Moore, Will H. 1995. "Rational Rebels: Overcoming the Free-rider Problem." *Political Research Quarterly* 48 (2): 417–454.

Moore, Jack. 2017. "ISIS Unleashes Dozens of Female Suicide Bombers in Battle for Mosul." *Newsweek*, July 5, 2017. http://www.newsweek.com/isis-female-suicide-bombers-battle-mosul-631846.

Mulrine, Anna. 2013. "8 Other Nations That Send Women to Combat." *National Geographic*, January 25, 2013. http://news.nationalgeographic.com/news/2013/13/130125-women-combat-world-australia-israel-canada-norway/.

Nacos, Brigitte. 2005. "The Portrayal of Female Terrorists in the Media: Similar Framing Patterns in the News Coverage of Women in Politics and Terrorism." *Studies in Conflict and Terrorism* 28 (5): 435–451.

Nagel, Joane. 1998. "Masculinity and Nationalism: Gender and Sexuality in the Making of Nations." *Ethnic and Racial Studies* 21 (2): 242–269.

Nash, Mary. 1993. "Women in the War: Milicianas and Armed Combat in Revolutionary Spain, 1936–1939." *International History Review* 15 (2): 269–282.

——. 1995. *Defying Male Civilization: Women in the Spanish Civil War.* Denver, CO: Arden Press.

Nepstad, Sharon. 2004. *Convictions of the Soul: Religious, Culture, and Agency in the Central American Solidarity Movement.* Cambridge: Cambridge University Press.

Nghidinwa, Marie. 2008. *Women Journalists in Namibia's Liberation Struggle, 1985–1990.* Basel, Switzerland: Basler Afrika Bibliographien.

Nhongo-Simbanegavi, Josephine. 2000. *For Better or Worse? Women and ZANLA in Zimbabwe's Liberation Struggle.* Harare, Zimbabwe: Weaver Press.

Öcalan, Abdullah. 2015a. "Democratic Modernity: Era of Woman's Revolution." http://pkkonline.org/en/index.php?sys=article&artID=235.

——. 2015b. "Killing the Dominant Male: Instituting the Third Major Sexual Rupture Against the Dominant Male." http://www.pkkonline.org/en/index.php?sys=article&artID=230..

O'Gorman, Eleanor. 2011. *The Frontline Runs Through Every Woman: Women and Local Resistance in the Zimbabwean Liberation War.* Harare, Zimbabwe: Weaver Press.

O'Keeffe, Alice. 2008. "Jungle Fever." *The Guardian*, August 23, 2008. https://www.theguardian.com/world/2008/aug/24/farc.colombia.

Oppel, Richard, Jr. 2015. "Two Female Soldiers Poised to Graduate from Ranger School." *New York Times*, August 17, 2015. http://www.nytimes.com/2015/08/18/us/two-female-soldiers-poised-to-graduate-from-ranger-school.html.

O'Rourke, Lindsey. 2009. "What's Special about Female Suicide Terrorism?" *Security Studies* 18: 681–718.

Özeren, Süleyman, Murat Sever, Kamil Yilmaz, and Alper Sözer. 2014. "Whom Do They Recruit? Profiling and Recruitment in the PKK/KCK." *Studies in Conflict and Terrorism* 37 (4): 322–347.

Parashar, Swati. 2011. "Gender, *Jihad*, and Jingoism: Women as Perpetrators, Planners, and Patrons of Militancy in Kashmir." *Studies in Conflict and Terrorism* 34 (4): 295–317.

Parkinson, Sarah. 2013. "Organizing Rebellion: Rethinking High-Risk Mobilization and Social Networks in War." *American Political Science Review* 107 (3): 418–432.

Pateman, Roy. 1990. *Eritrea: Even the Stones Are Burning.* Trenton, NJ: Red Sea Press.

Peek, Charles, George Lowe, and Susan Williams. 1991. "Gender and God's Word: Another Look at Religious Fundamentalism and Sexism." *Social Forces* 69 (4): 1205–1221.

Pennington, Reina. 2010. "Offensive Women: Women in Combat in the Red Army in the Second World War." *Journal of Military History* 74: 775–820.

Peralta, Eyder. 2015. "Marine Corps Study: All Male-Combat Units Performed Better Than Mixed Units." *NPR*, September 10, 2015. http://www.npr.org/sections/thetwo-way/2015/09/10/439190586/marine-corps-study-finds-all-male-combat-units-faster-than-mixed-units.

Phillips, David, and Kelly Berkell. 2016. "The Case for Delisting the PKK as a Foreign Terrorist Organization." *Lawfare*, February 11. https://www.lawfareblog.com/case-delisting-pkk-foreign-terrorist-organization.

Phippen, Weston. 2016. "Women on Britain's Frontlines." *The Atlantic*, July 8, 2016. https://www.theatlantic.com/news/archive/2016/07/british-military-women/490475/.

Piazza, James. 2009. "Is Islamist Terrorism More Dangerous? An Empirical Study of Group Ideology, Organization, and Goal Structure." *Terrorism and Political Violence* 21: 62–88.

PKK. 2019. "Homepage." January 24, 2019. http://www.pkkonline.org/en/index.php.

Poulos, Margaret. 2008. *Arms and the Woman: Just Warriors and Greek Feminist Identity*. New York: Columbia University Press.

Pugel, James. 2007. "What the Fighters Say: A Survey of Ex-Combatants in Liberia, February-March 2006." Monrovia, Liberia: United Nations Development Program Liberia.

Puts, David, Benedict Jones, and Lisa DeBruine. 2012. "Sexual Selection on Human Faces and Voices." *Journal of Sexual Research* 49 (2–3): 227–243.

Raghavan, S. V., and V. Balasubramaniyan. 2014. "Evolving Role of Women in Terror Groups: Progression or Regression." *Journal of International Women's Studies* 15 (2): 197–211.

Rao, Kathyayini, and Carol Tilt. 2016. "Board Composition and Corporate Responsibility: The Role of Diversity, Gender, Strategy and Decision Making." *Journal of Business Ethics* 138 (2): 327–347.

Reif, Linda. 1986. "Women in Latin American Guerrilla Movements: A Comparative Perspective." *Comparative Politics* 18 (2): 147–169.

Robinson, Kristopher, Edward Crenshaw, and J. Craig Jenkins. 2006. "Ideologies of Violence: The Social Origins of Islamist and Leftist Transnational Terrorism." *Social Forces* 84 (4): 2009–2026.

Robson, Seth. 2014. "Female Fighters of the PKK May Be the Islamic State's Worst Nightmare." *Stars and Stripes*, August 30, 2014. https://www.stripes.com/news

/female-fighters-of-the-pkk-may-be-the-islamic-state-s-worst-nightmare-1.300259#.WbG1end969Y.

Rodriquez, Fernando S., Theodore Curry, and Gang Lee. 2006. "Gender Differences in Criminal Sentencing: Do Effects Vary Across Violent, Property and Drug Offences?" *Social Science Quarterly* 87 (2): 318–338.

Ross, Tim. 2015. "British Army's Women Soldiers Go into Combat." *The Telegraph*, December 20, 2015. http://www.telegraph.co.uk/news/uknews/defence/12060225/British-Armys-women-soldiers-to-go-into-combat.html.

RT. 2015. "Her War: Women vs. ISIS." June 21. https://www.youtube.com/watch?v=uqIoa4VgEs8.

Rubin, Jacqueline, and Will H. Moore. 2007. "Risk Factors for Forced Migrant Flight." *Conflict Management and Peace Science* 24 (2): 85–104.

Ruddick, Sara. 1989. *Maternal Thinking: Towards a Politics of Peace*. Boston: Beacon Press.

Rudman, Lauri, and Stephanie Goodwin. 2004. "Gender Difference in Automatic In-Group Bias: Why Do Women Like Women More Than Men Like Men?" *Journal of Personality and Social Psychology* 87 (4): 494–509.

Rupp, Leila J. 1978. Mobilizing Women for War: German and American Propaganda, 1939-1945. Princeton, NJ: Princeton University Press.

Salehyan, Idean. 2009. *Rebels Without Borders: Transnational Insurgencies in World Politics*. Ithaca, NY: Cornell University Press.

Salehyan, Idean, and Kristian Skrede Gleditsch. 2006. "Refugees and the Spread of Civil War." *International Organization* 60 (2): 335–366.

Salehyan, Idean, Kristian Skrede Gleditsch, and David Cunningham. 2011. "Explaining External Support for Insurgent Groups." *International Organization* 65 (4): 709–744.

Salih, Mohammad. 2014. "Meet the Badass Women Fighting the Islamic State." Dispatch, *Foreign Policy*, September 12, 2014. http://foreignpolicy.com/2014/09/12/meet-the-badass-women-fighting-the-islamic-state./

Salopek, Paul. 2003. "Militia Tries Charm Offensive." *Chicago Tribune*, May 5, 2003. https://www.chicagotribune.com/news/ct-xpm-2003-05-05-0305050189-story.html.

San-Akca, Belgin. 2015. "Dangerous Companions: Cooperation Between States and Nonstate Armed Groups (NAGs)," v. 4. http://nonstatearmedgroups.ku.edu.tr/nag_list.php.

Santos, Soliman, Jr., and Paz Verdades M. Santos. 2010. *Primed and Purposeful: Armed Groups and Human Security Efforts in the Philippines*. Geneva, Switzerland: Small Arms Survey.

Sapire, Hilary. 2009. "Liberation Movements, Exile, and International Solidarity: An Introduction." *Journal of Southern African Studies* 35 (2): 271–286.

Saunders, Chris. 2009. "Namibian Solidarity: British Support for Namibian Independence." *Journal of Southern African Studies* 35 (2): 437–453.
Sawyer, Katherine, Kathleen Gallagher Cunningham, and William Reed. 2017. "The Role of External Support in Civil War Termination." *Journal of Conflict Resolution* 61 (6): 1174–1202.
Schalk, Peter. 1994. "Women Fighters of the Liberation Tigers in Tamil Ilam: The Martial Feminism of Atel Palacinkam." *South Asia Research* 14 (2): 163–195.
Schneidman, Witney. 1978. "FRELIMO's Foreign Policy and the Process of Liberation." *Africa Today* 25 (1): 57–67.
Schubiger, Livia Isabella, and Matthew Zelina. 2017. "Ideology in Armed Groups." *PS: Political Science and Politics* 50 (4): 948–952.
Schulze, Kristen. 2003. "The Struggle for an Independent Aceh: The Ideology, Capacity, and Strategy of GAM." *Studies in Conflict and Terrorism* 26 (4): 241–271.
Scott, James. 1976. *The Moral Economy of the Peasant: Rebellion and Subsistence in Southeast Asia.* New Haven, CT: Yale University Press.
Sellström, Tor. 2002. *Sweden and the National Liberation of Southern Africa.* Vol. 2, *Solidarity and Assistance, 1970–1994.* Uppsala, Sweden: Nordiska Afrikainstitutet.
Shair-Rosenfield, Sarah, and Reed M. Wood. 2017. "Governing Well after War: How Improving Female Representation Prolongs Post-Conflict Peace." *Journal of Politics* 79 (2): 995–1009.
Sharlach, Lisa. 1999. "Gender and Genocide in Rwanda: Women as Agents and Objects." *Journal of Genocide Research* 1 (3): 387–399.
Shehadeh, Lamia Rustum. 1999. "Women in the Lebanese Militias." In *Women and War in Lebanon*, ed. Lamia Rustum Shehadeh, 145–166. Gainseville: University of Florida Press.
Shekhawat, Seema. 2015. "Introduction: Women in Conflict and Peace-Making." In *Female Combatants in Conflict and Peace: Challenging Gender in Violence and Post-Conflict Reintegration*, ed. Seema Shekhawat, 1–19. New York: Palgrave MacMillan.
Shelton, Allison, Szymon M. Stojek, and Patricia L. Sullivan. 2013. "What Do We Know About Civil War Outcomes?" *International Studies Review* 15 (4): 515–538.
Sherkat, Darren, and Christopher Ellison. 1999. "Recent Developments and Current Controversies in the Sociology of Religion." *Annual Review of Sociology* 25: 363–394.
Shover, Michele. J. 1975. "Roles and Images of Women in World War I Propaganda." *Politics and Society* 5 (4): 469–486.
Sims, Helen. 2000. "Posters and Images of Women in the Great War." In *Representations of Gender from Prehistory to Present*, ed. Moira Donald and Linda Hurcombe, 168–181. New York: St. Martin's Press.

Singer, J. David, Stuart Bremer, and John Stuckey. 1972. "Capability Distribution, Uncertainty, and Major Power War, 1820–1965." In *Peace, War, and Numbers*, ed. Bruce Russett, 19–48. Beverly Hills, CA: Sage.

Singer, P. W. 2006. *Children at War.* Berkeley: University of California Press.

Sjoberg, Laura. 2006. "Gendered Realities and the Immunity Principle: Why Gender Analysis Needs Feminism." *International Studies Quarterly* 50 (4): 889–910.

——. 2010. "Women Fighters and the 'Beautiful Soul' Narrative." *International Review of the Red Cross* 92 (877): 53–68.

——. 2014. *Gender, War, and Conflict.* Cambridge: Polity Press.

——. 2016. *Women as Wartime Rapists: Beyond Sensation and Stereotyping.* New York: New York University Press.

——. 2018. "Jihadi Brides and Female Volunteers: Reading the Islamic State's War to See Gender and Agency in Conflict Dynamics." *Conflict Management and Peace Science* 35 (3): 296–311.

Sjoberg, Laura, and Caron Gentry. 2007. *Mothers, Monsters, Whores: Women's Violence in Global Politics.* New York: Zed Books.

Smith, Patrick. 2014. "The Kurdish 'Angels of Kobane' Are Fighting on a Second Front." *Newsweek*, December 11, 2014. http://www.newsweek.com/2014/12/19/angels-kobane-are-fighting-second-front-290835.html.

Speckhard, Anne. 2008. "The Emergence of Female Terrorism." *Studies in Conflict and Terrorism* 31 (11): 995–1023.

Spohn, Cassia, and Dawn Beichner. 2000. "Is Preferential Treatment of Female Offenders a Thing of the Past? A Multisite Study of Gender, Race and Imprisonment." *Criminal Justice Policy Review* 11 (2): 149–184.

Stack, Alisa. 2009. "Zombies Versus Black Widows: Women as Propaganda in the Chechen Conflict." In *Women, Gender and Terrorism*, ed. Laura Sjoberg and Caron Gentry, 83–95. Athens: University of Georgia Press.

Stack-O'Connor, Alisa. 2007a. "Lions, Tigers, and Freedom Birds: How and Why the Liberation Tigers of Tamil Eelam Employs Women." *Terrorism and Political Violence* 19 (1): 43–63.

——. 2007b. "Picked Last: Women and Terrorism." *Joint Forces Quarterly* 44 (1): 95–100.

Stam, Valerie. 2009. "Women's Agency and Collective Action: Peace Politics in the Casamance." *Canadian Journal of African Studies* 43 (2): 337–366.

Staniland, Paul. 2014. *Networks of Rebellion: Explaining the Politics of Insurgent Cohesion and Collapse.* Ithaca, NY: Cornell University Press.

Stanski, Keith. 2006. "Terrorism, Gender, and Ideology: A Case Study of Women Who Join the Revolutionary Armed Forces of Colombia (FARC)." In *The Making of a Terrorist: Recruitment, Training, and Toot Causes*, ed. James J. F. Forest, 136–150. Westport, CT: Praeger Security International.

START (National Consortium for the Study of Terrorism and Responses to Terrorism). 2016. Global Terrorism Database. Data file. https://www.start.umd.edu/gtd.

Stokke, Kristian. 2006. "Building the Tamil Eelam State: Emerging State Institutions and Forms of Governance in LTTE-Controlled Areas in Sri Lanka." *Third World Quarterly* 27 (6): 1021–1040.

Stott, Leda. 1990. "Women and the Armed Struggle for Independence in Zimbabwe (1964–1979)." Occasional Papers, no. 25. Edinburgh University, Center for African Studies.

Subedi, D. B. 2013. "From Civilian to Combatant: Armed Recruitment and Participation in the Maoist Conflict in Nepal." *Contemporary South Asia* 21 (4): 429–443.

Sullivan, John. 1988. *ETA and Basque Nationalism: The Fight for Euskadi, 1890–1986.* London: Routledge.

Tan, Mine Gögüs. 2007. "Women, Education, and Development in Turkey." In *Education in "Multicultural" Societies: Turkish and Swedish Perspectives*, ed. Marie Carlson, Annika Rabo, and Fatma Gök, 107–122. Stockholm: Swedish Research Institute in Istanbul.

Tareke, Gebru. 2009. *The Ethiopian Revolution: War in the Horn of Africa.* New Haven, CT: Yale University Press.

Tavakolian, Newsha. 2015. "Meet the Women Taking the Battle to ISIS." *Time.* April 2, 2015. http://time.com/3767133/meet-the-women-taking-the-battle-to-isis/.

Taylor, Sandra. 1999. *Vietnamese Women at War: Fighting for Ho Chi Minh and the Revolution.* Lawrence: University of Kansas Press.

Tezcür, Günes. 2015. "Violence and Nationalist Mobilization: The Onset of the Kurdish Insurgency in Turkey." *Nationalities Papers* 43 (2): 248–266.

Thomas, Jakana, and Kanisha Bond. 2015. "Women's Participation in Violent Political Organizations." *American Political Science Review* 109 (3): 488–506.

Thomas, Jakana, and Reed M. Wood. 2018. "The Social Origins of Female Combatants." *Conflict Management and Peace Science* 35 (3): 215–232.

Thompson, Carol B. 1982. "Women in the Liberation Struggle in Zimbabwe." *Women's Studies International Forum* 5 (4): 247–252.

Toivanen, Mari, and Bahar Baser. 2016. "Gender in the Representation of an Armed Conflict: Female Kurdish Combatants in the French and British Media." *Middle East Journal of Culture and Communication* 9: 294–314.

Tovar, Juan Camilo Maldonoado. 2015. "Guerrillas in the Mist: Seven Days in Rebel-Held Territory in Colombia." *VICE*, December 3, 2015. http://www.vice.com/read/guerrillas-in-the-mist-v22n12.

Turner, Karen. 1998. *Even the Women Must Fight: Memories of War from North Vietnam.* With Phan Thanh Hao. New York: John Wiley and Sons.

Twum-Danso, Afua. 2003. *Africa's Young Soldiers: The Co-optation of Childhood.* Monograph 82, April. Pretoria, South Africa: Institute for Security Studies.

Ugarriza, Juan, and Matthew Craig. 2012. "The Relevance of Ideology to Contemporary Armed Conflicts: A Quantitative Analysis of Former Combatants in Colombia." *Journal of Conflict Resolution* 57 (3): 445–477.

UK Ministry of Defense. 2014. "Women in Ground Close Combat (GCC) Review Paper." December 19. https://www.gov.uk/government/publications/women-in-ground-close-combat-gcc-review-paper.

UN Women. 2012. *Identifying Women's and Peace and Security Priorities.* http://www.unwomen.org/~/media/Headquarters/Media/Publications/en/03BIdentifyingWomens.pdf.

Uppsala Conflict Data Program (UCDP). 2018. *UCDP Conflict Encyclopedia.* http://ucdp.uu.se/.

Urdang, Stephanie. 1984. "Women in National Liberation Movements." In *African Women South of the Sahara,* ed. Margaret Jean Hay and Sharon Stichter, 156–169. London: Longman.

Utas, Mats. 2005. "Victimcy, Girlfriending, Soldiering: Tactical Agency in a Young Woman's Social Navigation of the Liberian Conflict Zone." *Anthropological Quarterly* 78 (2): 403–430.

Van Bruinessen, Martin. 2001. "From Adela Khanum to Leyla Zana: Women as Political Leaders in Kurdish History." In *Women of a Non-state Nation: The Kurds,* ed. Shahrzad Mojab, 95–112. Costa Mesa, CA: Mazda Publishers.

Van Evera, Stephen. 1994. "Hypotheses on Nationalism and War." *International Security* 18 (4): 5–39.

Varghese, Johnlee. 2015. "Rehana, the Kurdish Female Fighter Who 'Killed' 100 ISIS Fighters, Spotted in Kobane." *International Business Times,* June 8, 2015. http://www.ibtimes.co.in/rehana-kurdish-female-fighter-who-killed-100-isis-fighters-spotted-kobane-photo-635123.

Veale, Angela. 2003. *From Child Soldier to Ex-Fighter: Female Fighters, Demobilisation and Integration in Ethiopia.* Pretoria, South Africa: Institute for Security Studies.

VICELAND. 2016. "Colombia: The Women of FARC." Episode 2 of *WOMAN with Gloria Steinem.* Aired May 17, on VICELAND.

VICE News. 2012. "Female Fighters of Kurdistan." July 23, 2012. https://www.youtube.com/watch?v=h_okg8VlxkE.

Viterna, Jocelyn. 2006. "Pulled, Pushed, and Persuaded: Explaining Women's Mobilization into the Salvadoran Guerrilla Army." *American Journal of Sociology* 112 (1): 1–45.

——. 2013. *Women in War: The Micro-processes of Mobilization in El Salvador.* Oxford: Oxford University Press.

———. 2014. "Radical or Righteous? Using Gender to Shape Public Perceptions of Political Violence." In *Dynamics of Political Violence: A Process-Oriented Perspective on Radicalization and the Escalation of Political Conflict*, ed. Charles Demetriou, Lorenzo Bosi, and Stefan Malthaner, 189–216. Burlington, VT: Ashgate.

Vogel, Lauren, Louise Porter, and Mark Kebbell. 2014. "The Roles of Women in Contemporary Political and Revolutionary Conflict: A Thematic Model." *Studies in Conflict and Terrorism* 37: 91–114.

Von Knop, Katharina. 2007. "The Female Jihad: Al-Qaeda's Women." *Studies in Conflict and Terrorism* 30: 397–414.

Wallace, Michael, and J. Craig Jenkins. 1995. "The New Class, Postindustrialism, Neocorporatism: Three Images of Social Protest in the Western Democracies." In *The Politics of Social Protest: Comparative Perspectives on States and Social Movements*, ed. J. Craig Jenkins and Bert Klandermans, 96–137. Minneapolis: University of Minnesota Press.

Walter, Barbara. 2009. "Bargaining Failures in Civil War." *Annual Review of Political Science* 12: 243–261.

Walters, Joanna. 2015."'Flawed' Study Casts Doubts on Mixed-Gender Units in US Marine Corps." *The Guardian*, October 17, 2015. https://www.theguardian.com/us-news/2015/oct/17/marines-study-casts-doubts-mixed-gender-units.

Wayland, Sarah. 2004. "Ethnonationalist Networks and Transnational Opportunities: The Sri Lanka Tamil Diaspora." *Review of International Studies* 30 (3): 405–426.

Weinstein, Jeremy M. 2007. *Inside Rebellion: The Politics of Insurgent Violence*. Cambridge: Cambridge University Press.

Weiss, Ruth. 1986. *The Women of Zimbabwe*. London, UK: Kesho Publishers.

Wells, Julia C. 2003. "The Sabotage of the Patriarchy in Rural Rhodesia: Rural African Women's Living Legacy to Their Daughters." *Feminist Review* 75: 101–117.

White, Paul. 2015. *The PKK: Coming Down from the Mountains*. London: Zed Books.

Wickham-Crowley, Timothy. 1992. *Guerrillas and Revolution in Latin America: A Comparative Study of Insurgents and Regimes since 1956*. Princeton, NJ: Princeton University Press.

Williams, John E., and Deborah L. Best. 1990. *Measuring Sex Stereotypes: A Multination Study*. Newbury Park, CA: Sage.

Williams, Kristin. 2015. "Women in Armed Groups Are More Than Just an Exotic Novelty." *Public Radio International*, June 28, 2015. http://www.pri.org/stories/2015-06-28/women-armed-groups-are-more-just-exotic-novelty.

Winter, Charlie. 2015. *Women of the Islamic State: A Manifesto on Women by the Al-Khanssaa Brigade*. Quilliam Foundation, February. https://therinjfoundation.files.wordpress.com/2015/01/women-of-the-islamic-state3.pdf.

Wood, Elisabeth Jean. 2003. *Insurgent Collective Action in El Salvador.* Cambridge: Cambridge University Press.

——. 2006. "Variation in Sexual Violence During War." *Politics and Society* 34 (3): 307–341.

——. 2008. "The Social Processes of Civil War: The Wartime Transformation of Social Networks." *Annual Review of Political Science* 11: 539–561.

——. 2009. "Armed Groups and Sexual Violence: When Is Wartime Rape Rare?" *Politics and Society* 37 (1): 131–162.

Wood, Elisabeth Jean, and Nathaniel Toppelberg. 2017. "The Persistence of Sexual Assault within the US Military." *Journal of Peace Research* 54 (5): 620–633.

Wood, Reed M. 2014. "From Loss to Looting? Battlefield Costs and Rebel Incentives for Violence." *International Organization* 68 (4): 979–999.

Wood, Reed M., and Jakana Thomas. 2017. "Women on the Frontline: Rebel Group Ideology and Women's Participation in Violent Rebellion." *Journal of Peace Research* 54 (1): 31–46.

Wood, Wendy, and Alice Eagly. 2002. "A Cross-Cultural Analysis of the Behavior of Women and Men: Implications for the Origins of Sex Differences." *Psychological Bulletin* 128 (5): 699–727.

Woodward, Rachel, and Patricia Winter. 2004. "Discourses of Gender in the Contemporary British Army." *Armed Forces and Society* 30 (2): 279–301.

Woolf, Virginia. (1938) 1963. *Three Guineas.* Orlando, FL: Harvest Books.

World Bank. 2018. "World Development Indicators." Washington, DC. http://data.worldbank.org/data-catalog/world-development-indicators.

Yarchi, Moran. 2014. "The Effect of Female Suicide Attacks on Foreign Media Framing of Conflicts: The Case of Palestinian-Israeli Conflict." *Studies in Conflict and Terrorism* 37: 674–688.

YJA-Star. 2019. "Our Martyrs." yja-star.com. January 24, 2019. http://www.yja-star.com/ku/sehitlerimiz.

YPG. 2017. "When Female Fighters Lead the Charge." ypgrojava.org. July 28, 2017. https://www.ypgrojava.org/When-Female-Fighters-Lead-the-Charge.

Yuan, Lijun. 2005. *Reconceiving Women's Equality in China: A Critical Examination of Models of Sex Equality.* Lanham, MD: Lexington Books.

Young, John. 1997. *Peasant Revolution in Ethiopia.* Cambridge: Cambridge University Press.

Yücesahin, Murat, and Murat Özgür. 2008. "Regional Fertility Differences in Turkey: Persistent High Fertility in the Southeast." *Population, Space and Place* 14: 135–155.

Yüksel, Metin. 2006. "The Encounter of Kurdish Women with Nationalism in Turkey." *Middle Eastern Studies* 42 (5): 777–802.

Yuval-Davis, Nira. 1993. "Gender and Nation." *Ethnic and Racial Studies* 16 (4): 621–632.

Zelditch, Morris. 2001. "Theories of Legitimacy." In *The Psychology of Legitimacy*, ed. John Jost and Brenda Major, 33–54. Cambridge: Cambridge University Press.

Zunes, Stephen, and Jacob Mundy. 2010. *Western Sahara: War, Nationalism, and Conflict Irresolution*. Syracuse, NY: Syracuse University Press.

Index

Note: Page numbers followed by *f* and *t* refer to figures and tables respectively.